QUANTUM FIELD THEORY AND THE MANY-BODY PROBLEM

The Many-Body Problem

Current Research and Reviews

Edited by Irwin Oppenheim, Department of Chemistry, Massachusetts Institute of Technology

Volume I

T. D. Schultz, QUANTUM FIELD THEORY AND THE MANY-BODY PROBLEM

Volume II

E. Meeron, Ed., PHYSICS OF MANY-PARTICLE SYSTEMS: METHODS AND PROBLEMS

Additional volumes in preparation

QUANTUM FIELD THEORY AND THE MANY-BODY PROBLEM

T. D. SCHULTZ

Thomas J. Watson Research Center
I.B.M. Corporation
Yorktown Heights, N.Y.

GORDON AND BREACH, SCIENCE PUBLISHERS
NEW YORK . LONDON . PARIS

GORDON AND BREACH, Science Publishers, Inc.
150 Fifth Avenue, New York, N.Y. 10011
Library of Congress Catalog Card Number: 64-17898

Editorial office for Great Britain:
Gordon and Breach, Science Publishers Ltd.
8 Bloomsbury Way
London W.C.1

Editorial Office for France:
Gordon & Breach
7-9 rue Emile Dubois
Paris 14^e

Distributed in Canada by:
The Ryerson Press
299 Queen Street West
Toronto 2B, Ontario

Printed in the Republic of Ireland

PREFACE

This is a set of notes based on two series of lectures. The first, a series of six lectures under the auspices of the Greater Washington Solid State Physics Colloquium and the National Bureau of Standards, was given at the Bureau of Standards in June, 1960. The second series, somewhat enlarged, was given at the Physical Research Laboratory of Space Technology Laboratories, Inc. during the summer, 1960. The notes, prepared by the author, first appeared as a technical report of Space Technology Laboratories in October, 1960.

The intention of the lectures was to provide an introduction to the recent developments in the theory of the many-body problem, which depend heavily on the formalism of quantum field theory—an introduction that would be accessible to those hardly acquainted with quantum field theory.

The scope of the lectures was necessarily limited by considerations of time. For this reason, I have been content to present only the perturbation theory approach via diagrams, rather than a broader and less accessible survey. These lectures are then to be considered a detailed introduction to only one aspect of the modern developments in the many-body problem.

In the spring of 1962, when the notes were prepared for publication, certain revisions and extensions were made and some attempt was also made to include a few references to more recent work where it was particularly pertinent. During this period I profited considerably from conversations with colleagues in the mathematical physics group at the Thomas J. Watson Research Center of I.B.M. I am particularly indebted to Dr. Arthur Sakakura of this group, to Prof. Harvey Kaplan of Syracuse University, and to Drs. Lewis Klein and Walter Sadowski of the National Bureau of Standards, for many valuable suggestions. I am also very grateful to Dr. Burton Fried for his hospitality at Space Technology Laboratories and to my colleagues there for many stimulating discussions.

CONTENTS

INTRODUCTION

Many of the problems in solid-state physics are inherently many-body problems, although it has often been customary to idealize these problems so that one has to deal with only a few particles in interaction, or many particles interacting with each other only in some averaged way. In recent years there has been considerable interest, and a great number of papers to show for it, in trying to treat the many-particle aspects of solid-state physics problems in some more nearly exact way. In particular, the emphasis has often been on those aspects of these problems which involve correlated motions of many particles, and so lie completely outside any one-particle treatment. In these attempts, several widely diverse ideas and approaches have been used, too many certainly to describe in this brief set of lectures. Because of limitations of time, I shall restrict my consideration mainly to one approach, the many-body perturbation theory and the analysis using Feynman diagrams. Thus, I shall have to exclude consideration of two of the most interesting problems these days, those of liquid helium and superconductivity, because the many-body perturbation theory is not *directly* applicable to either of these. The perturbation approach has the advantages, however, of being systematic and understandable in physical terms, both of which seem suitable for an introduction. It also turns out that the formalism and techniques of the perturbation theory seem to have application beyond their apparent validity. One can even treat the problems of liquid helium and superconductivity with the perturbation theory formalism, if one first makes the appropriate modifications.[1,2]

The many-body perturbation theory is really the perturbation theory of quantum field theory, which has been studied in great detail over the past ten or twelve years.* Still, many of the results for the

* See any textbook on quantum field theory like Jauch and Rohrlich, Akhiezer and Berestetsky, or Schweber, Bethe and de Hoffman.

many-body problem have been obtained in ways not directly following the main line of quantum field theory. I shall try to get these results in the most direct and concise way, even though this might not do justice to the recent history in this subject. For this purpose I won't assume anything beyond elementary quantum mechanics.

References

1. N. M. Hugenholtz and D. Pines, *Phys. Rev.* **116**, 489 (1959).
2. Y. Nambu, *Phys. Rev.* **117**, 648 (1960).

CHAPTER I

SECOND QUANTIZATION

Consider a system of N electrons in a volume Ω. Suppose it has a Hamiltonian of the form

$$H = \sum_{i=1}^{N} \frac{\mathbf{p}_i^2}{2m} + \sum_{i=1}^{N} w(\mathbf{r}_i) + \tfrac{1}{2} \sum_{i,j}{}' v(\mathbf{r}_i - \mathbf{r}_j). \tag{1.1}$$

Here $w(\mathbf{r})$ is a one-body potential (say, an externally applied potential like that due to the field of the ions in a solid), and $v(\mathbf{r}_i - \mathbf{r}_j)$ is a two-body potential like the Coulomb potential between electrons. $\sum'_{i,j}$ means to sum over all i and j except $i = j$.

We wish to introduce the formalism of second quantization. This is not absolutely *necessary*, because the number of particles in the system described by the Hamiltonian in (1.1) is obviously fixed; one need never consider the creation or destruction of particles. The second quantization formalism is introduced solely for certain conveniences:

(a) It allows us to introduce an electron-hole picture where our "particles" are either electrons or holes. There are then often only a few "particles" although these are not conserved in number. It may be much simpler to talk of one or two particles than always to talk about the 10^{22} electrons in a solid.

(b) In quantum statistical mechanics, it is convenient to work with the Grand Canonical Ensemble, in which case we must simultaneously consider systems of different numbers of particles. For this too, the second quantization formalism is very convenient.

For the electron-hole system in solids, I must emphasize that second quantization gives no new physics. It is just a new notation with much greater freedom than the customary first quantization formalism. This situation is different from quantum electrodynamics, where the second quantization of the electron-positron field really is a new physical idea.

Consider, for our system, a complete orthonormal set of one-particle functions

$$u_1(\mathbf{r}), \quad u_2(\mathbf{r}), \ldots$$

obeying some convenient set of boundary conditions (e.g. periodic boundary conditions), and assume that these functions are arranged in some definite order. Then it is possible to construct a complete set of totally antisymmetric n-body wave functions ("Slater determinants")

$$\sum_{\mathscr{P}}(-1)^p \mathscr{P} u_{i_1}(\mathbf{r}_1) \ldots u_{i_n}(\mathbf{r}_n), \tag{1.2}$$

where $(i_1 \ldots i_n)$ is *any* n-tuple of integers such that $i_1 < i_2 < \ldots < i_n$. $\mathscr{P}$ permutes the sub-subscripts so that

$$\mathscr{P} u_{i_1}(\mathbf{r}_1) \ldots u_{i_n}(\mathbf{r}_n) = u_{i_{\mathscr{P}1}}(\mathbf{r}_1) \ldots u_{i_{\mathscr{P}n}}(\mathbf{r}_n),$$

and p is the number of interchanges of pairs of sub-subscripts necessary to get from the old to the new order.

I have implicitly left spin out of this. To include it in our discussion is very easy. We would just have to think of $\mathbf{r}$ as a symbol representing a vector position *and* a spin orientation (up or down), too. Let us just assume our particles have no spin (although they obey Fermi statistics) and then later we shall mention how to put it back in.

Now let us introduce an abstract Hilbert space, the vectors of which represent possible states of our system and are in a one-to-one correspondence with the linear combinations of the functions (1.2). For a system of n particles, we would have one Hilbert space and, for a system of n' particles, we would have another, etc. Let us consider one great Hilbert space obtained by combining all these. A complete set of basis vectors in this space would be the following:

- Φ_0, representing a system with no particles at all (bare vacuum),
- Φ_i, all i, representing a system with one particle with wave function $u_i(\mathbf{r})$,
- $\Phi_{i_1 i_2}$, all (i_1, i_2) such that $i_1 < i_2$, representing a system with two particles described by the wave function

$$\sum_{\mathscr{P}}(-)^p\mathscr{P}u_{i_1}(\mathbf{r}_1)u_{i_2}(\mathbf{r}_2), \tag{1.3}$$

and so forth.

Note that $\Phi_{i_2i_1}$ will correspond to

$$\sum_{\mathscr{P}}(-1)^p\mathscr{P}u_{i_2}(\mathbf{r}_1)u_{i_1}(\mathbf{r}_2)$$

and so $\Phi_{i_1i_2} = -\Phi_{i_2i_1}$, etc.

We can now define a set of linear operators $C_i^\dagger$ which "create" a particle in the states $u_i(\mathbf{r})$ in the following sense:

$$C_i^\dagger\Phi_{i_1\ldots i_n} = \Phi_{ii_1\ldots i_n}. \tag{1.4}$$

Since this defines what the $C_i^\dagger$ do to all the basis vectors in the space, it completely defines these operators. From this definition we can now prove several simple properties of these "creation operators":

(1) $$\Phi_{i_1\ldots i_n} = C_{i_1}^\dagger \ldots C_{i_n}^\dagger\Phi_0, \tag{1.5a}$$

obvious by induction.

(2) $$C_i^\dagger C_j^\dagger = -C_j^\dagger C_i^\dagger. \tag{1.5b}$$

Proof: $C_i^\dagger C_j^\dagger\Phi_{i_1\ldots i_n}$ has the wave function

$$\sum_{\mathscr{P}}(-1)^p\mathscr{P}u_i(\mathbf{r}_{n+2})u_j(\mathbf{r}_{n+1})u_1(\mathbf{r}_1)\ldots u_n(\mathbf{r}_n)$$

and $C_j^\dagger C_i^\dagger\Phi_{i_1\ldots i_n}$ has the wave function

$$\sum_{\mathscr{P}}(-1)^p\mathscr{P}u_j(\mathbf{r}_{n+2})u_i(\mathbf{r}_{n+1})u_1(\mathbf{r}_1)\ldots u_n(\mathbf{r}_n).$$

The two wave functions are the same but of opposite sign. Since this is true for any basis vector in the space, it is true for any vector and the theorem is proved.

A corollary:

$$C_i^{\dagger 2}=0 \tag{1.5b'}$$

(3) Let us now define the adjoint operators C_i. The notation seems a little odd, taking as fundamental the operators with the

†'s and as their adjoints, operators without the †'s. Still, this convention is almost universal so we'll follow it. To define the C_i explicitly, we must see how they act on any basis vector. Consider the vector $C_i\Phi_{i_1 \ldots i_s}$ which we expand in a complete set of basis vectors:

$$C_i\Phi_{i_1 \ldots i_s} = \sum_{t,(i_1' \ldots i_t')} \langle C_i\Phi_{i_1 \ldots i_s}|\Phi_{i_1' \ldots i_t'}\rangle \Phi_{i_1' \ldots i_t'}.$$

A typical coefficient is

$$\begin{aligned}\langle C_i\Phi_{i_1 \ldots i_s}|C_{i_1'}{}^\dagger \ldots C_{i_t'}{}^\dagger\Phi_0\rangle &= \langle \Phi_{i_1 \ldots i_s}|C_i{}^\dagger C_{i_1'}{}^\dagger \ldots C_{i_t'}{}^\dagger\Phi_0\rangle \\ &= 0 \text{ unless the set } (i_1 \ldots i_s) \text{ equals} \\ &\quad\text{the set } (i, i_1' \ldots i_t'). \\ &= (-1)^{r-1} \text{ if } i_1' = i_1, \text{etc., and } i = i_r.\end{aligned}$$

We thus conclude that C_i destroys a particle in the ith one-particle state or annihilates the state $\Phi_{i_1 \ldots i_s}$ altogether, if it contains no particle in the ith state.

(4) From (1.5b), taking adjoints, we have

$$C_iC_j = -C_jC_i; \quad C_i^2 = 0. \tag{1.5c}$$

(5) We can now prove another important result:

$$C_iC_j{}^\dagger = -C_j{}^\dagger C_i \quad \text{if} \quad i \neq j$$

and (1.5d)

$$C_iC_i{}^\dagger + C_i{}^\dagger C_i = 1.$$

Proof: If $i \neq j$

$$\begin{aligned}C_iC_j{}^\dagger\Phi_{i_1 \ldots i_s} &= (-)^r\Phi_{ji_1 \ldots \not{r} \ldots i_s} \quad \text{if} \quad i = i_r, \\ &= 0 \text{ if } i \text{ not in } (j, i_1 \ldots i_s); \\ C_j{}^\dagger C_i\Phi_{i_1 \ldots i_s} &= (-)^{r-1}\Phi_{ji_1 \ldots \not{r} \ldots i_s} \quad \text{if} \quad i = i_r, \\ &= 0 \text{ if } i \text{ not in } (i_1 \ldots i_s).\end{aligned}$$

Thus, if $i \neq j$, either

$$C_iC_j{}^\dagger\Phi_{i_1 \ldots i_s} = -C_j{}^\dagger C_i\Phi_{i_1 \ldots i_s}$$

or

$$C_i C_j^\dagger \Phi_{i_1 \ldots i_s} = C_j^\dagger C_i \Phi_{i_1 \ldots i_s} = 0,$$

so that

$$C_i C_j^\dagger = -C_j^\dagger C_i.$$

And if $i = j$,

$$\begin{aligned} C_i C_i^\dagger \Phi_{i_1 \ldots i_s} &= \Phi_{i_1 \ldots i_s} \quad \text{if } i \text{ is not in } (i_1 \ldots i_s), \\ &= 0 \text{ if } i \text{ is in } (i_1 \ldots i_s); \\ C_i^\dagger C_i \Phi_{i_1 \ldots i_s} &= 0 \text{ if } i \text{ is not in } (i_1 \ldots i_s), \\ &= \Phi_{i_1 \ldots i_s} \text{ if } i \text{ is in } (i_1 \ldots i_s). \end{aligned}$$

Adding, we see that

$$(C_i C_i^\dagger + C_i^\dagger C_i) \Phi_{i_1 \ldots i_s} = \Phi_{i_1 \ldots i_s}$$

or

$$C_i C_i^\dagger + C_i^\dagger C_i = 1.$$

Observe that the operator $C_i^\dagger C_i$, giving zero if the ith one-particle state is not occupied and one if it is, can be thought of as the occupation number operator for this state.

We may summarize these properties, using the notation $\{A, B\}$ for $AB + BA$, the anticommutator:

$$\begin{aligned} \{C_i, C_j\} &= \{C_i^\dagger, C_j^\dagger\} = 0, \\ \{C_i, C_j^\dagger\} &= \delta_{ij}. \end{aligned} \tag{1.5}$$

So far we have just introduced a Hilbert space, a set of basis vectors, and some linear operators which map basis vectors into other basis vectors (or annihilate them). Now comes the crucial step: for any operator in ordinary first quantization, there is a corresponding operator in the Hilbert space whose matrix elements between any pair of basis vectors in the Hilbert space are exactly the same as the matrix elements of the first quantization operator between the corresponding Slater determinants. Specifically, one can verify the following correspondences (setting $\hbar = 1$ for convenience):

$$\sum_i w(\mathbf{r}_i) \rightleftharpoons \sum_{\alpha\beta} \langle \alpha | w | \beta \rangle C_\alpha^\dagger C_\beta, \tag{1.6a}$$

where

$$\langle\alpha|w|\beta\rangle = \int d^3r\, u_\alpha^*(\mathbf{r})w(\mathbf{r})u_\beta(\mathbf{r});$$

$$\sum_i \frac{\nabla_i^2}{2m} \rightleftharpoons \sum_{\alpha\beta}\left\langle\alpha\left|\frac{\nabla^2}{2m}\right|\beta\right\rangle C_\alpha^\dagger C_\beta, \tag{1.6b}$$

where

$$\left\langle\alpha\left|\frac{\nabla^2}{2m}\right|\beta\right\rangle = \int d^3r\, u_\alpha^*(\mathbf{r})\frac{\nabla^2}{2\mathrm{m}}u_\beta(\mathbf{r});$$

and

$$\tfrac{1}{2}\sum_{ij}{}' v(\mathbf{r}_i-\mathbf{r}_j) \rightleftharpoons \tfrac{1}{2}\sum_{\alpha\beta\gamma\delta}\langle\alpha\beta|v|\gamma\delta\rangle C_\alpha^\dagger C_\beta^\dagger C_\gamma C_\delta, \tag{1.6c}$$

where

$$\langle\alpha\beta|v|\gamma\delta\rangle = \int d^3r\int d^3r'\, u_\alpha^*(\mathbf{r})u_\beta^*(\mathbf{r}')v(\mathbf{r}-\mathbf{r}')u_\gamma(\mathbf{r}')u_\delta(\mathbf{r}).$$

If u_α, u_β, u_γ, u_δ are plane wave functions $e^{i\mathbf{k}_\alpha\cdot\mathbf{r}}/\Omega^{\frac{1}{2}}$, etc., and Ω is allowed to become extremely large, then

$$H = \sum_p \frac{\mathbf{p}^2}{2m}C_p^\dagger C_p + \frac{1}{\Omega}\sum_{p,q} w(\mathbf{q})C_p^\dagger C_{p-q} + \frac{1}{2\Omega}\sum_{pp'q} v(\mathbf{q})C_p^\dagger C_{p'}^\dagger C_{p'+q}C_{p-q},$$

where

$$w(\mathbf{q}) = \int d^3r\, e^{-i\mathbf{p}\cdot\mathbf{r}}\, w(\mathbf{r})\, e^{i(\mathbf{p}-\mathbf{q})\cdot\mathbf{r}} = \int d^3r\, e^{-i\mathbf{q}\cdot\mathbf{r}}\, w(\mathbf{r}),$$

and

$$v(\mathbf{q}) = \int d^3r\, e^{-i\mathbf{q}\cdot\mathbf{r}}\, v(\mathbf{r}).$$

It is convenient to introduce operators $\bar{\psi}(\mathbf{r})$ and $\psi(\mathbf{r})$ which are linear combinations of the creation and annihilation operators:

$$\bar{\psi}(\mathbf{r}) = \Omega^{-\frac{1}{2}}\sum_p C_p^\dagger\, e^{-i\mathbf{p}\cdot\mathbf{r}}; \quad \psi(\mathbf{r}) = \Omega^{-\frac{1}{2}}\sum_p C_p\, e^{i\mathbf{p}\cdot\mathbf{r}}. \tag{1.8}$$

Just as $C_p^\dagger$ and C_p create and destroy a particle in the momentum state $\mathbf{p}$, so $\bar{\psi}(\mathbf{r})$ and $\psi(\mathbf{r})$ create and destroy a particle at $\mathbf{r}$ ($C_p^\dagger$ puts a particle in the one-particle state $e^{i\mathbf{p}\cdot\mathbf{r}}$; so $\bar{\psi}(\mathbf{r})$ puts a particle in the one-particle state $\sum_p e^{i\mathbf{p}\cdot(\mathbf{r}'-\mathbf{r})} \propto \delta(\mathbf{r}-\mathbf{r}')$).

In terms of $\bar{\psi}$ and ψ we have

$$H = \int d^3r\, \bar{\psi}(\mathbf{r})\left(-\frac{\nabla^2}{2m}+w(\mathbf{r})\right)\psi(\mathbf{r})$$
$$+\tfrac{1}{2}\int d^3r\, d^3r'\, v(\mathbf{r}-\mathbf{r}')\bar{\psi}(\mathbf{r})\bar{\psi}(\mathbf{r}')\psi(\mathbf{r}')\psi(\mathbf{r}). \tag{1.9}$$

To interpret an operator like $\nabla^2\psi(\mathbf{r})$ we must expand according to (1.8) and then operate with ∇^2 on each coefficient.

From the definitions (1.8) and the properties of the C_p and $C_p{}^\dagger$ (1.5), it follows immediately that, for $\Omega \to \infty$,

$$\{\psi(\mathbf{r}),\ \psi(\mathbf{r}')\} = \{\bar{\psi}(\mathbf{r}),\ \bar{\psi}(\mathbf{r}')\} = 0, \tag{1.10}$$
$$\{\psi(\mathbf{r}),\ \bar{\psi}(\mathbf{r}')\} = \delta(\mathbf{r}-\mathbf{r}').$$

It is convenient to rewrite (1.9) so that each pair of creation and annihilation operators referring to the same point in space stand together. Then the two-body term in H becomes

$$\tfrac{1}{2}\int d^3r\, d^3r'\, v(\mathbf{r}-\mathbf{r}')\bar{\psi}(\mathbf{r})\psi(\mathbf{r})\bar{\psi}(\mathbf{r}')\psi(\mathbf{r}')-\tfrac{1}{2}v(\mathbf{r}=0)\int d^3r\, \bar{\psi}(\mathbf{r})\psi(\mathbf{r}).$$

The second term, which is just $-\frac{1}{2}Nv(\mathbf{r}=0)$, can be considered to redefine the zero of the single particle energy and can be included in the kinetic energy part of H. This addition and subtraction of the energy of interaction of each particle with itself unfortunately has the disadvantage that $v(\mathbf{r}=0)$ may be infinite (as it is for the Coulomb interaction—we'll refer to $\frac{1}{2}Nv(\mathbf{r}=0)$ as the "infinite self-energy"). This is not a fundamental difficulty; we can introduce a cutoff at some value of $\mathbf{r}$ (or equivalently, for some large value of $\mathbf{q}$ in the Fourier transform of $v(\mathbf{r})$) and then let this cutoff go to zero after *all* calculations have been made. No real problem ever arises in doing this if care is taken, because the terms in $v(\mathbf{r}=0)$ must always cancel. This is discussed further in sections G1, H1 and H2 of Chapter III.

It will also usually be convenient to assume that $w(\mathbf{r})$ contains a "background" part, $w_b(\mathbf{r})$, that cancels a part of the two-body term. For the case of a Coulomb two-body interaction, this would be the potential of a uniform background of positive charge equal and opposite in charge density to the average charge density of the electrons. To make this notion more precise, consider the Fourier transform of the particle density operator:

$$\rho(\mathbf{q}) = \int d^3r\, e^{-i\mathbf{q}\cdot\mathbf{r}}\, \bar{\psi}(\mathbf{r})\, \psi(\mathbf{r}) = \sum_p C_p{}^\dagger C_{p+q}.$$

2

In terms of this operator, the potential energy part of H (apart from the self-energy term) is written*

$$H_1 = \Omega^{-1} \sum_q [w(\mathbf{q})\rho(\mathbf{q}) + \tfrac{1}{2}v(\mathbf{q})\rho(\mathbf{q})\rho(-\mathbf{q})],$$

$$\underset{\Omega\to\infty}{=} \int \frac{d^3q}{(2\pi)^3}[w(\mathbf{q})\rho(\mathbf{q}) + \tfrac{1}{2}v(\mathbf{q})\rho(\mathbf{q})\rho(-\mathbf{q})].$$

Now $\rho(\mathbf{q} = 0)$ is just the operator for the total number of particles. It commutes with H and can be assumed to have the c-number value N. It contributes to H_1 the sum

$$\frac{1}{\Omega}[Nw(\mathbf{q} = 0) + \tfrac{1}{2}N^2 v(\mathbf{q} = 0)].$$

Rather than carry this constant in H_1, it is convenient to consider it an additive constant in the total energy, which can be defined away by a proper choice of the zero of energy,† and to assume that H_1 has no $\mathbf{q} = 0$ contribution. This is guaranteed if we set $v(\mathbf{q} = 0) = w(\mathbf{q} = 0) = 0$ in all that follows. This will have no effect except in expressions like $\int d^3q\, v(\mathbf{q})\delta(\mathbf{q})$ which now vanish.

* In transforming the two-body term, one must make use of the fact that the potential $v(\mathbf{r})$ is not long range. This insures that a negligible error is made when an integral over the variables $\mathbf{r}_1$ and $\mathbf{r}_2$ is written as an integral over $\mathbf{r}_1 - \mathbf{r}_2$ and $\frac{1}{2}(\mathbf{r}_1 + \mathbf{r}_2)$ with limits on the two new variables taken to be independent of one another. The assumption that $v(\mathbf{r})$ has finite range is equivalent to the requirement that $v(\mathbf{q})$ remain finite as $\mathbf{q} \to 0$.

† In the Coulomb case, this is not really necessary. The uniform positive background gives rise to $w(\mathbf{q} = 0) = -Nv(\mathbf{q} = 0)$. It also contributes a term to the total energy resulting from the interaction of the positive charge distribution with itself $= \frac{1}{2}N^2 v(\mathbf{q} = 0)$. The sum of the three interaction energies between uniform charge distributions (background–background, electron–electron, and electron–background) is then zero. This is discussed in some detail at the beginning of Chapter IV.

CHAPTER II

HEISENBERG AND SCHRÖDINGER PICTURES

Let us just recall that there are two customary ways of describing the time development of a system quantum mechanically, the Schrödinger and Heisenberg pictures.

Schrödinger Picture

The operators (C_p's or ψ's) are time independent.

The state vector is time dependent and satisfies the Schrödinger equation

$$i\frac{\partial \Psi^S(t)}{\partial t} = H\Psi^S(t). \tag{2.1}$$

This has the formal solution (if H does not depend explicitly on the time):

$$\Psi^S(t) = e^{-iHt}\Psi^S(0). \tag{2.1$'$}$$

Heisenberg Picture

The state vector is time independent. Define it to agree with the Schrödinger state vector at $t = 0$:

$$\Psi^H = \Psi^S(0)$$

The operators change with time, in such a way that the matrix element of an operator between two states in the Heisenberg picture is the same as the matrix element of the corresponding operator between the corresponding two states in the Schrödinger picture:

$$\langle \Psi_a{}^S(t)|\, \mathcal{O}^S|\Psi_b{}^S(t)\rangle = \langle \Psi_a{}^H|\, \mathcal{O}^H(t)|\Psi_b{}^H\rangle.$$

Using (2.1′), we have

$$\langle \Psi_a{}^S(0)|\exp(iHt)\, \mathcal{O}^S \exp(-iHt)|\Psi_b{}^S(0)\rangle = \langle \Psi_a{}^H|\, \mathcal{O}^H(t)|\Psi_b{}^H\rangle,$$

so that

$$\mathcal{O}^H(t) = e^{iHt}\, \mathcal{O}^S\, e^{-iHt}. \tag{2.3$'$}$$

If we apply the operator $i\partial/\partial t$ to (2.3′)

$$i\frac{\partial \mathcal{O}^H(t)}{\partial t} = i\, e^{iHt}\, (iH\, \mathcal{O}^S - \mathcal{O}^S iH)\, e^{-iHt}.$$

$$= i^2\, (H\, e^{iHt}\, \mathcal{O}^S\, e^{-iHt} - e^{iHt}\, \mathcal{O}^S\, e^{-iHt}\, H),$$

we see that (2.3′) is the formal solution of the Heisenberg equation of motion

$$i\frac{\partial \mathcal{O}^H(t)}{\partial t} = [\, \mathcal{O}^H(t), H].$$

We have been rather careless about saying whether the Hamiltonian is in the Heisenberg or Schrödinger picture. Actually, for systems where H does not depend explicitly on time $H^S = H^H$, i.e., the Hamiltonian is a constant of the motion. H may still be expressed with either Schrödinger or Heisenberg operators, and although the total Hamiltonian is the same in the two pictures, various parts of it may be quite different in the two pictures.

CHAPTER III

MANY-FERMION SYSTEMS AT ABSOLUTE ZERO

A. Introduction and Definition of Green Functions

Let us first consider a system of many identical fermions. We have in mind the conduction electrons in a metal, or the nucleons in a nucleus (neglecting spin and isotopic spin). Among the many properties we should like to know about such a system are the following: the ground state energy, the nature of the elementary excitations, the equilibrium thermodynamic properties, and the behavior in applied fields. A systematic way to investigate all of these properties is through the various Green functions of the system, as we shall see.

The use of Green functions has several advantages. First, we can avoid some divergence difficulties encountered in ordinary perturbation theory when applied to very large systems (e.g., ordinary perturbation theory gives contributions to the ground state energy proportional to *all* positive integral powers of the system volume, which we ultimately wish to be arbitrarily large). This is essential, because these divergences are certainly not physical but rather are the fault of the mathematics of ordinary perturbation theory. Second, the approach via Green functions does not depend on the use of perturbation theory, a particular advantage when perturbation theory breaks down, as for example, in the case of superconductivity. In these notes, we shall however keep pretty much to a perturbational approach to Green functions, although the results have greater generality than the approach itself implies. A third advantage of the Green function approach, is that most of the quantities of immediate physical interest can be expressed in terms of the so-called one- and two-particle Green functions. It is true that, for some properties, the Green functions contain more information about the system than is needed, but on the other hand these two Green functions contain far less unnecessary information than the full set of n-particle stationary states of the system. A fourth advantage of the Green function approach, is that there has been extensive development of this formalism for its applications in quantum electrodynamics and other quantum field theories. Thus you will find much of what we shall discuss in the textbooks already cited.

I should say another word at the start: not all of the recent activity on the many-body problem has been developed along the lines of Green functions. In particular the work of Van Hove, Hugenholtz, Goldstone, Bloch and de Dominicis, Watson *et al.*, Thouless, and others, though closely related to the development I am going to give is a little different.

The *n-particle Green function in space-time variables* is defined by

$$G(x_1 \dots x_n;\quad x_1' \dots x_n')$$
$$= (-i)^n \langle \Psi_0 | T(\psi(x_1) \dots \psi(x_n) \bar{\psi}(x_n') \dots \bar{\psi}(x_1')) | \Psi_0 \rangle, \qquad (3.1)$$

where $x = (\mathbf{r}, t)$, Ψ_0 is the exact ground state, Heisenberg picture, and $\psi(x) = \psi(\mathbf{r}, t)$ and $\bar{\psi}(x) = \bar{\psi}(\mathbf{r}, t)$ are Heisenberg operators destroying (and creating) a particle at the point $\mathbf{r}$ at time t. By T we mean that the operators following T should be rearranged from the order in which they stand on the paper so that (1) those with later values of their time variable stand to the left of those with earlier time variables, (2) for operators with the same value of their time variables, the $\bar{\psi}$ operators stand to the left of the ψ operators, and (3) for each permutation of two adjacent operators needed to bring the operators into the desired order from the order on the paper, a factor (-1) is introduced. Such a product is called a T-product.

An example of a Green function is the one-particle function

$$G(x, x') = -i\langle \Psi_0 | \psi(x) \bar{\psi}(x') | \Psi_0 \rangle \quad \text{if} \quad t > t',$$
$$= +i\langle \Psi_0 | \bar{\psi}(x') \psi(x) | \Psi_0 \rangle \quad \text{if} \quad t < t'.$$

Actually because $G(x, x')$ is a discontinuous function of t at $t = t'$, the definition at this special point is really arbitrary. On pp. 61–2, it will be shown that for some purposes it should be the arithmetic mean of the limiting values when $t \to t'$ from each side.

Green functions are also called propagators. There is in the literature no uniformity, unfortunately, in the inclusion of factors of $(\pm i)$ in the definitions, so one must be a bit careful. I will discuss the meaning of Green functions in greater detail a little later.

It is often convenient to work with various Fourier transforms of $G(x, x')$ rather than $G(x, x')$ itself. If we introduce the expansions (1.8) in the definition of $G(x, x')$ we have

$$G(x, x') = -i\Omega^{-1} \sum_{p, p'} e^{i(\mathbf{p}\cdot\mathbf{r} - \mathbf{p}'\cdot\mathbf{r}')} \langle \Psi_0 | T(C_p(t)C_{p'}^{\dagger}(t')) | \Psi_0 \rangle, \qquad (3.2)$$

which suggests we define another Green function

$$G(\mathbf{p}t, \mathbf{p}'t') = -i\langle \Psi_0 | T(C_p(t)C_{p'}^{\dagger}(t')) | \Psi_0 \rangle. \qquad (3.3)$$

For time-independent systems, the Green function in (3.3) will depend only on the time difference $(t-t')$ and not on t and t' separately. To see this we note simply that

$$\begin{aligned} &\langle \Psi_0 | T(C_p(t)C_{p'}^{\dagger}(t')) | \Psi_0 \rangle \\ &\quad = \langle \Psi_0 | e^{iHt'}\, T(C_p(t-t')\, C_{p'}^{\dagger}(0))\, e^{-iHt'} | \Psi_0 \rangle \\ &\quad = \langle \Psi_0 | T(C_p(t-t')C_{p'}^{\dagger}(0)) | \Psi_0 \rangle\,. \end{aligned}$$

Now it is easy to show that the operator $C_p(t)$ decreases the total momentum of the system by $\mathbf{p}$ for all times (i.e., if $\mathscr{P}(t)\Psi = \mathbf{P}\Psi$ where $\mathscr{P}(t)$ is the total momentum operator at time t and equals $\Sigma \mathbf{p} C_p^{\dagger}(t)C_p(t)$, then $\mathscr{P}(t)C_p(t)\Psi = (\mathbf{P}-\mathbf{p})C_p(t)\Psi$). Similarly, $C_p^{\dagger}(t)$ increases the momentum by $\mathbf{p}$ for all times. For *homogeneous systems,* for which the Hamiltonian commutes with $\mathscr{P}$, the ground state will be an eigenstate of $\mathscr{P}$ with momentum zero, so that the matrix element in (3.3) will be zero unless $\mathbf{p} = \mathbf{p}'$. This suggests defining a Green function for homogenous time-independent systems

$$G(\mathbf{p}, t) = -i\langle \Psi_0 | T(C_p(t)C_p^{\dagger}(0)) | \Psi_0 \rangle. \qquad (3.4)$$

For such systems $G(x, x')$ is a function only of $x-x'$, call it $G(x-x')$. Then

$$G(x) = \Omega^{-1} \sum_{p} e^{i\mathbf{p}\cdot\mathbf{r}}\; G(\mathbf{p}, t).$$

In the limit that $\Omega \to \infty$ and the allowed values of $\mathbf{p}$ become quasi-continuous, the summation becomes an integral, so that

$$G(x) = \int \frac{d^3p}{(2\pi)^3} G(\mathbf{p}, t)\, e^{i\mathbf{p}\cdot\mathbf{r}} \qquad (3.5a)$$

and

$$G(\mathbf{p}, t) = \int d^3r \, G(x) \, e^{-i\mathbf{p} \cdot \mathbf{r}}. \tag{3.5b}$$

It is often convenient to introduce the four-dimensional Fourier transform $G(\mathbf{p}, E) \equiv G(p)$:

$$G(p) = \int dt \, e^{iEt} \, G(\mathbf{p}, t) = \int d^4x \, e^{-ip \cdot x} \, G(x), \tag{3.6a}$$

where

$$p \equiv (\mathbf{p}, E)$$

and

$$p \cdot x \equiv \mathbf{p} \cdot \mathbf{r} - Et.$$

Then

$$G(x) = \int \frac{d^4p}{(2\pi)^4} e^{ip \cdot x} \, G(p). \tag{3.6b}$$

It may happen that integrals on E encounter singularities in the integrand. In such a case, the path of integration must be carefully defined. I shall return to this question later.

B. Motivation to Study Green Functions

Before going into the detailed nature of the Green functions and the methods for calculating them for a given system, I should like to spend a few moments showing what kinds of detailed information we can expect to find about a system once we know its various Green functions.

First, the operator for the density of particles at the point x in space-time is $\bar{\psi}(x)\psi(x)$ so that the average density of particles at x is

$$\rho(x) = \langle \Psi_0 | \bar{\psi}(x)\psi(x) | \Psi_0 \rangle = -iG(\mathbf{r}t, \mathbf{r}t+\delta). \tag{3.7}$$

δ is a positive infinitesimal. Similarly, for homogeneous time-independent systems, the number of particles per unit volume in a small volume element of momentum space about the momentum $\mathbf{p}$ is

$$n(\mathbf{p}) = \frac{\Omega}{(2\pi)^3} \langle \Psi_0 | C_p^{\dagger}(0) C_p(-\delta) | \Psi_0 \rangle = -i \frac{\Omega}{(2\pi)^3} G(\mathbf{p}, t = -\delta)$$

$$= -i \frac{\Omega}{(2\pi)^3} \int_{-\infty}^{\infty} \frac{dE}{2\pi} e^{iE\delta} \, G(p).$$

We can close the contour with a great circle in the upper half-plane, giving

$$n(\mathbf{p}) = -i\frac{\Omega}{(2\pi)^3} \int_{\curvearrowleft} \frac{dE}{2\pi} e^{iE\delta}\, G(p). \tag{3.7'}$$

The "kinetic energy" of a system with Hamiltonian (1.9) and the modifications discussed at the end of Chapter I is

$$KE = \left\langle \Psi_0 \middle| \int d^3r\, \bar{\psi}(\mathbf{r}t)\left(-\frac{\nabla^2}{2m} - \tfrac{1}{2}v(\mathbf{r}=0)\right)\psi(\mathbf{r}t) \middle| \Psi_0 \right\rangle$$

$$= \int d^3r \left[i\frac{\nabla_r{}^2}{2m} G(\mathbf{r}t, \mathbf{r}'t+\delta) \right]_{r'=r} - \tfrac{1}{2}Nv(\mathbf{r}=0). \tag{3.8a}$$

For a homogeneous time-independent system, this can be further simplified:

$$KE = -i\Omega \int \frac{d^4p}{(2\pi)^4} \frac{p^2}{2m} e^{iE\delta}\, G(p) - \tfrac{1}{2}Nv(\mathbf{r}=0)$$

$$= \int d^3p \frac{p^2}{2m} n(\mathbf{p}) - \tfrac{1}{2}Nv(\mathbf{r}=0). \tag{3.8a'}$$

Similarly, the potential energy due to the one-body potential $w(\mathbf{r})$ is

$$PE_1 = -i \int w(\mathbf{r}) G(\mathbf{r}t, \mathbf{r}t+\delta)\, d^3r. \tag{3.8b}$$

For a homogeneous time-independent system, $w(\mathbf{r}) = w$, a constant, so that (3.8b) simplifies to

$$PE_1 = wN. \tag{3.8b'}$$

The potential energy due to the two-body interaction $v(\mathbf{r}-\mathbf{r}')$ when the infinite self-energy has been added, is

$$PE_2 = \left\langle \Psi_0 \middle| \tfrac{1}{2} \int \bar{\psi}(\mathbf{r}t)\psi(\mathbf{r}t)v(\mathbf{r}-\mathbf{r}')\bar{\psi}(\mathbf{r}'t)\psi(\mathbf{r}'t)\, d^3r\, d^3r' \middle| \Psi_0 \right\rangle.$$

This is simply written in terms of a Green function:

$$PE_2 = -\tfrac{1}{2} \int G(\mathbf{r}t, \mathbf{r}'t'; \mathbf{r}t^+, \mathbf{r}'t'^+) v(\mathbf{r}-\mathbf{r}') \delta(t-t'^{++}) \delta(t)\, d^4x\, d^4x', \tag{3.8c}$$

where t^+ means a time infinitesimally greater than t, and t^{++} is a time infinitesimally greater than t^+.

It is possible to determine the total energy of the state Ψ_0 knowing only the potential energy PE_2, the self-energy $-\frac{1}{2}Nv(\mathbf{r}=0)$, and the energy of the system of non-interacting particles. To see this, suppose we consider our Hamiltonian to have the form

$$H = T+\lambda V, \quad \text{(ultimately,} \quad \lambda = 1)$$

and suppose $\mathscr{E}_0(\lambda)$ and $\Psi_0(\lambda)$ are defined by the eigenvalue equation

$$H\Psi_0 = \mathscr{E}_0\Psi_0.$$

Then

$$\mathscr{E}_0 = \langle\Psi_0|H|\Psi_0\rangle,$$

$$\frac{\partial\mathscr{E}_0}{\partial\lambda} = \left\langle\frac{\partial\Psi_0}{\partial\lambda}\middle|H\middle|\Psi_0\right\rangle+\left\langle\Psi_0\middle|H\middle|\frac{\partial\Psi_0}{\partial\lambda}\right\rangle+\left\langle\Psi_0\middle|\frac{\partial H}{\partial\lambda}\middle|\Psi_0\right\rangle$$

$$= \mathscr{E}_0\frac{\partial}{\partial\lambda}\langle\Psi_0|\Psi_0\rangle+\left\langle\Psi_0\middle|\frac{\partial H}{\partial\lambda}\middle|\Psi_0\right\rangle.$$

But

$$\left\langle\frac{\partial\Psi_0}{\partial\lambda}\middle|\Psi_0\right\rangle+\left\langle\Psi_0\middle|\frac{\partial\Psi_0}{\partial\lambda}\right\rangle = \frac{\partial}{\partial\lambda}\langle\Psi_0|\Psi_0\rangle = 0,$$

since we can take $\langle\Psi_0|\Psi_0\rangle = 1$ for all λ. Thus

$$\frac{\partial\mathscr{E}_0}{\partial\lambda} = \left\langle\Psi_0\middle|\frac{\partial H}{\partial\lambda}\middle|\Psi_0\right\rangle,$$

often called "Feynman's Theorem". Integrating,

$$\mathscr{E}_0(\lambda = 1) = \mathscr{E}_0(\lambda = 0)+\int_0^1\frac{d\lambda}{\lambda}\langle\Psi_0|\lambda V|\Psi_0\rangle, \tag{3.9}$$

a relation that has been attributed to Pauli. For the free electron gas $(w(\mathbf{r}) \equiv 0)$, $E_0(\lambda = 0)$ is just the energy of the free Fermi sea, so we need only calculate $\langle\Psi_0|\lambda V|\Psi_0\rangle$, for all values of the electron charge between 0 and the true electron charge. Thus the problem of determining the ground state energy of an electron gas is reduced to determining PE_2 or just $G(\mathbf{r}t, \mathbf{r}'t'; \mathbf{r}t^+, \mathbf{r}'t'^+)$. We shall see much later that (making use of the equation of motion for $\psi(\mathbf{r}t)$) this Green function for these special values of the arguments can be expressed in terms of the one-particle Green function, i.e., $G(\mathbf{r}t, \mathbf{r}'t')$ itself is also sufficient to determine the ground state energy.

The response to external electric and magnetic fields can also be expressed in terms of Green functions, which we shall discuss a bit in connection with the dielectric constant.[1]

C. Free One-Particle Green Functions

1. *Starting with the Bare Vacuum*

Suppose we take Ψ_0 to be the state of no particles whatsoever, the "bare vacuum" state. Then, we have for various Green functions for $t < 0$:

$$G(x) = i\langle\Psi_0|\bar{\psi}(0)\psi(x)|\Psi_0\rangle = 0$$

$$G(\mathbf{p}, t) = 0;$$

and for $t > 0$:

$$G(\mathbf{p}, t) = -i\langle\Psi_0|C_p(t)C_p^\dagger(0)\Psi_0\rangle;$$

but since

$$C_p(t) = e^{iHt}\, C_p(0)\, e^{-iHt} = e^{-i\epsilon(\mathbf{p})t}\, C_p,$$

we have

$$G(\mathbf{p}, t) = -\, i\, e^{-i\epsilon(\mathbf{p})t}$$

and

$$G(x) = -\, i \int \frac{d^3p}{(2\pi)^3} e^{i\mathbf{p}\cdot\mathbf{r}}\, e^{-i\epsilon(\mathbf{p})t}.$$

$G(x)$ is the wave function of particle known to be at the origin at $t = 0$. $\epsilon(\mathbf{p})$ is the one-particle energy, which for the case of completely free particles ($w(\mathbf{r}) = 0$) is $p^2/2m$. We can combine both the cases of $t < 0$ and $t > 0$ in $G(p)$:

$$G(p) = \int_{-\infty}^{\infty} dt\, e^{iEt}\, G(\mathbf{p}, t) = -i \int_0^{\infty} dt\, e^{iEt}\, e^{-i\epsilon(\mathbf{p})t}.$$

This is well defined only for E above the real axis in which case

$$G(p) = \frac{1}{E-\epsilon(\mathbf{p})}.$$

We can allow E to be on the real axis if we add a small imaginary term to the denominator:

$$G(p) = \frac{1}{E-\epsilon(\mathbf{p})+i\delta},$$

where E is on or above the real axis, and δ is a positive infinitesimal.

2. *Starting with the Free Fermi Vacuum*

Suppose Ψ_0 to be the free Fermi sea, i.e., a single Slater determinant using all one-particle states up to the Fermi momentum p_F, which is determined by the number of particles in the system (we assume spherical symmetry for $\epsilon(\mathbf{p})$). Then

$$C_p{}^\dagger\Psi_0 = 0 \quad \text{for} \quad |\mathbf{p}| < p_F,$$
$$C_p\Psi_0 = 0 \quad \text{for} \quad |\mathbf{p}| > p_F.$$

Now it is easy to show that (we shall call these Green functions G_0, since they are a zeroth order approximation when interactions are to be included by perturbation theory)
<u>for $t > 0$:</u>

$$\begin{aligned} G_0(\mathbf{p}, t) &= 0 \quad \text{if} \quad |\mathbf{p}| < p_F \\ &= -i\, e^{-i\epsilon(\mathbf{p})t} \quad \text{if} \quad |\mathbf{p}| > p_F, \end{aligned} \tag{3.10a}$$

and

$$G_0(x) = -i \int_{|\mathbf{p}|>p_F} \frac{d^3p}{(2\pi)^3} e^{i(\mathbf{p}\cdot\mathbf{r}-\epsilon(\mathbf{p})t)}; \tag{3.10b}$$

<u>and for $t < 0$:</u>

$$\begin{aligned} G_0(\mathbf{p}, t) &= 0 \quad \text{if} \quad |\mathbf{p}| > p_F \\ &= i\, e^{-i\epsilon(\mathbf{p})t} \quad \text{if} \quad |\mathbf{p}| < p_F \end{aligned} \tag{3.10a$'$}$$

and

$$G_0(x) = \int_{|\mathbf{p}|<p_F} \frac{d^3p}{(2\pi)^3} e^{i(\mathbf{p}\cdot\mathbf{r}-\epsilon(\mathbf{p})t)}. \tag{3.10b$'$}$$

Thus, for $t = 0$, if we try to add one particle at the origin, we are partly frustrated, since there is a finite probability of already finding a particle there. $G(x)$ is thus not precisely a wave function of the system with one additional particle known to start at the origin. Rather, it is the probability amplitude that, knowing a particle to be at the origin at $t = 0$ (we cannot distinguish between the one we add and all those of the Fermi sea), we find one at $\mathbf{r}$ at time t. Similarly, for $t < 0$, we create a hole (i.e., we annihilate a particle) at $\mathbf{r}$ at time t and $G(x)$ is the probability amplitude that there is no particle (i.e., that there is a hole) at the origin at $t = 0$.

It is most useful to compute $G_0(p)$. For $|\mathbf{p}| > p_F$, just as in the case of the bare vacuum, we find

$$G_0(p) = \frac{1}{E - \epsilon(\mathbf{p})}, \text{ where } E \text{ must be above the real axis,}$$

or

$$G_0(p) = \frac{1}{E - \epsilon(\mathbf{p}) + i\delta}, \ E \text{ on or above the real axis.}$$

Similarly, for $|\mathbf{p}| > p_F$

$$G_0(p) = \frac{1}{E - \epsilon(\mathbf{p}) - i\delta}, \ E \text{ on or below the real axis.}$$

We may combine these two results into

$$G_0(p) \equiv G_0(\mathbf{p}, E) = \frac{1}{E - \epsilon(\mathbf{p}) + i\delta_p}, \ E \text{ real,} \tag{3.11}$$

where

$$\begin{aligned} \delta_p &= \delta \quad \text{if} \quad |\mathbf{p}| > p_F \\ &= -\delta \quad \text{if} \quad |\mathbf{p}| < p_F. \end{aligned}$$

3. *Starting with a Fermi Sea of Bloch Waves*

The results above are not really restricted to the case of a free Fermi sea. If we have a nonzero one-body potential $w(\mathbf{r})$, which may even be a self-consistent potential added to the unperturbed part of the Hamiltonian and subtracted from the interaction part, then it may be convenient to expand the operators $\psi(x)$ and $\bar{\psi}(x)$ in Bloch waves.

If the $u_k(\mathbf{r})$'s are an orthonormal set of Bloch functions in an extended zone scheme, we can define operators which create and destroy particles in these one-particle states:

$$\xi(\mathbf{k}, t) = \int d^3r\, u_k^*(\mathbf{r})\psi(x), \quad \xi^\dagger(\mathbf{k}, t) = \int d^3r\, u_k(\mathbf{r})\bar{\psi}(x);$$

$$\psi(x) = \sum u_k(\mathbf{r})\xi(\mathbf{k}, t), \qquad \bar{\psi}(x) = \sum u_k^*(\mathbf{r})\xi^\dagger(\mathbf{k}, t).$$

All the above expressions are then still exact if

(a) we replace C_p and $C_p{}^\dagger$ by ξ_k and $\xi_k{}^\dagger$,

(b) we replace $\exp(i\mathbf{p}\,.\,\mathbf{r})/\Omega^{1/2}$ by $u_k(\mathbf{r})$, and

(c) we interpret the $\epsilon(\mathbf{k})$ to be the one-electron energies in the potential $w(\mathbf{r})$.

One important difference should be noted. For interacting "free" electrons, the total momentum $\mathscr{P}$ is conserved. For interacting Bloch electrons, the total Bloch wave vector

$$\mathscr{K} = \sum \mathbf{k}\xi_k{}^\dagger\xi_k$$

is not necessarily conserved. This means that we can't define a simple Green function like $G(\mathbf{p}, t)$ but rather must consider a more general function $G(\mathbf{k}, \mathbf{k}', t)$ in which $\mathbf{k}-\mathbf{k}'$ may be any vector of the reciprocal lattice times 2π.

D. Exact One-Particle Green Functions

It is possible to deduce a number of properties of the exact one-particle Green functions from their definitions. In addition to being useful formal properties, they will give us some insight into the physical meaning of these functions. In this section, I shall follow quite closely the paper of Galitskii and Migdal.[2]

1. *Spectral Representation*

For noninteracting particles we found a very convenient form for $G_0(p)$, viz.

$$G_0(p) = \frac{1}{E-\epsilon(\mathbf{p})+i\delta_p}, \quad p = (\mathbf{p}, E).$$

Let us now try to find a generalization of this expression which holds for the exact $G(p)$. An easy approach is to consider $G(\mathbf{p}, t)$.

For $t > 0$:

$$G(\mathbf{p}, t) = -i \langle \Psi_0 | C_p(t) C_p{}^\dagger(0) | \Psi_0 \rangle$$
$$= -i \langle \Psi_0 | e^{iHt} C_p e^{-iHt} C_p{}^\dagger | \Psi_0 \rangle.$$

Now $\exp(iHt)$, acting to the left, gives a factor $\exp(i\mathscr{E}_0{}^N t)$, where $\mathscr{E}_0{}^N$ is the ground state energy of an N-particle system. Let us expand the remaining matrix element in a complete set of intermediate states, energy states for the $N+1$ particle problem which we shall call $\Psi_n{}^{N+1}$ (since the operators $C_p{}^\dagger$ and C_p only connect Ψ_0 with $N+1$ particle states when Ψ_0 is to the right of $C_p{}^\dagger$ or to the left of C_p).

$$G(\mathbf{p}, t) = -i \sum_{nn'} e^{i\mathscr{E}_0{}^N t} \langle \Psi_0 | C_p | \Psi_n{}^{N+1} \rangle$$
$$\langle \Psi_n{}^{N+1} | e^{-iHt} | \Psi_{n'}{}^{N+1} \rangle \langle \Psi_{n'}{}^{N+1} | C_p{}^\dagger | \Psi_0 \rangle.$$

But

$$\langle \Psi_n{}^{N+1} | e^{-iHt} | \Psi_{n'}{}^{N+1} \rangle = e^{-i\mathscr{E}_n{}^{N+1} t} \delta_{nn'},$$

so in an obvious notation,

$$G(\mathbf{p}, t) = -i \sum_n |(C_p{}^\dagger)_{n0}|^2 \exp\{-i(\mathscr{E}_n{}^{N+1} - \mathscr{E}_0{}^N)t\} \quad \text{for} \quad t > 0,$$

the sum index n going over all stationary states of $N+1$ particles (and we need only include states of total momentum $\mathbf{p}$ since Ψ_0 is presumed to have momentum zero).

The energy difference $\mathscr{E}_n{}^{N+1} - \mathscr{E}_0{}^N$ really consists of two parts, the energy needed to go from the ground state of N particles to the ground state of $N+1$, and the energy needed to excite an $N+1$ particle system from its ground state to its nth state:

$$\mathscr{E}_n{}^{N+1} - \mathscr{E}_0{}^N = (\mathscr{E}_0{}^{N+1} - \mathscr{E}_0{}^N) + (\mathscr{E}_n{}^{N+1} - \mathscr{E}_0{}^{N+1}).$$

In the first parentheses is just μ^N, the chemical potential for an N-particle system; in the second parentheses is just $\epsilon_n{}^{N+1}$, an exact excitation energy of an $N+1$ particle system. Thus

$$G(\mathbf{p}, t) = -i \sum_n |(C_p{}^\dagger)_{n0}|^2 \exp\{-i(\epsilon_n{}^{N+1} + \mu^N)t\}, \qquad t > 0. \qquad (3.12a)$$

Similarly, for $t < 0$:

$$G(\mathbf{p}, t) = i \sum_n |(C_p)_{n0}|^2 \exp\{i(\epsilon_n{}^{N-1} - \mu^{N-1})t\}, \qquad t < 0, \tag{3.12b}$$

where n runs over states of $N-1$ particles, momentum $-\mathbf{p}$.

Henceforth we shall neglect the difference $\mu^N - \mu^{N-1}$ as being of order $1/N$. This would be appropriate for a metal but *not* for an insulator with just filled bands.

Two things should be observed:

(a) $\epsilon_n{}^{N+1}$, $\epsilon_n{}^{N-1}$ are always positive energies, being excitation energies.

(b) C_p (and similarly, $C_p{}^\dagger$) when applied to the exact ground state Ψ_0 creates a linear combination of states with a spread of energies.

Let us define $\rho^+(\mathbf{p}, \epsilon)d\epsilon$ to be the probability that $C_p{}^\dagger\Psi_0$ has its excitation energy in $(\epsilon, \epsilon + d\epsilon)$:

$$\rho^+(\mathbf{p}, \epsilon)d\epsilon \equiv \sum_n |(C_p{}^\dagger)_{n0}|^2, \qquad \epsilon < \epsilon_n{}^{N+1} < \epsilon + d\epsilon. \tag{3.13a}$$

Similarly, define $\rho^-(\mathbf{p}, \epsilon)$:

$$\rho^-(\mathbf{p}, \epsilon)d\epsilon \equiv \sum_n |(C_p)_{n0}|^2, \qquad \epsilon < \epsilon_n{}^{N-1} < \epsilon + d\epsilon. \tag{3.13b}$$

Because of the first remark above,

$$\rho^+(\mathbf{p}, \epsilon) = \rho^-(\mathbf{p}, \epsilon) = 0 \quad \text{for} \quad \epsilon < 0. \tag{3.14}$$

With the functions ρ^+ and ρ^- (which, of course, we do not know in general), we can convert the sums in (3.12a, b) into integrals. We can also make the Fourier transformation on the time variable obtaining

$$G(p) = \int_0^\infty dt\, e^{iEt}\left\{-i\int d\epsilon\, \rho^+(\mathbf{p}, \epsilon)\, e^{-i(\epsilon+\mu)t}\right\} + \int_{-\infty}^0 dt\, e^{iEt}\left\{i\int d\epsilon\, \rho^-(\mathbf{p}, \epsilon)\, e^{i(\epsilon-\mu)t}\right\}.$$

Let us now interchange orders of integration. The resulting time integrals are similar to those encountered for $G_0(p)$. We obtain

$$G(p) = \int_0^\infty d\epsilon\left\{\frac{\rho^+(\mathbf{p}, \epsilon)}{E-(\mu+\epsilon)+i\eta} + \frac{\rho^-(\mathbf{p}, \epsilon)}{E-(\mu-\epsilon)-i\eta}\right\}. \tag{3.15}$$

This is Lehmann's spectral representation for $G(p)$ which is exact and from which we can deduce various analytic properties and relations for $G(p)$.

The form (3.15) is very similar to that found for the noninteracting system. There $C_p^\dagger$ raised the energy by exactly $\epsilon(\mathbf{p})$ (if $|\mathbf{p}| > p_F$), corresponding to an excitation energy $\epsilon(\mathbf{p})-\mu$, so that $\rho^+(\mathbf{p}, \epsilon) = \delta[\epsilon-(\epsilon(\mathbf{p})-\mu)](1-n_p)$, the factor $(1-n_p)$ being unity for $|\mathbf{p}| > p_F$ and zero for $|\mathbf{p}| < p_F$. Similarly C_p lowered the energy $\epsilon(\mathbf{p})$ corresponding to an excitation energy $\mu-\epsilon(\mathbf{p})$ (if $|\mathbf{p}| < p_F$) so that $\rho^-(\mathbf{p}, \epsilon) = \delta[\epsilon-(\mu-\epsilon(\mathbf{p})]n_p$. For nearly free particles, we might expect $\rho^+(\mathbf{p}, \epsilon)$ and $\rho^-(\mathbf{p}, \epsilon)$ to be approximations to these delta functions.

2. *Singularity of $G(\mathbf{p}, E)$ for $E \to \mu$*

There is a well-known relation

$$\frac{1}{x \pm i\eta} = \mathscr{P}\frac{1}{x} \mp \pi i\delta(x), \eta \text{ a positive infinitesimal,}$$

the interpretation being that, if we multiply by any reasonable function $f(x)$ and integrate from $-\infty$ to ∞, interpreting $\mathscr{P}$ to mean principal value, the relation is borne out.

The integrals in (3.15) are of this form. Let us take the imaginary part of both sides of (3.15). Since ρ^+ and ρ^- are both positive real quantities, the delta function terms contribute on the right and we have

$$Im G(\mathbf{p}, E) = -\pi\rho^+(\mathbf{p}, E-\mu)+\pi\rho^-(\mathbf{p}, \mu-E). \tag{3.17}$$

Recalling that $\rho^+(\mathbf{p}, \epsilon) = \rho^-(\mathbf{p}, \epsilon) = 0$ for $\epsilon < 0$, we have

$$\begin{aligned} Im G(\mathbf{p}, E) &= -\pi\rho^+(\mathbf{p}, E-\mu), \quad E > \mu, \\ &= +\pi\rho^-(\mathbf{p}, \mu-E), \quad E < \mu. \end{aligned} \tag{3.18}$$

Thus, as E moves along the real axis (for fixed $\mathbf{p}$), the imaginary part of $G(\mathbf{p}, E)$ has a discontinuity at $E = \mu$.

3. *A Dispersion Relation for $Re G(\mathbf{p}, E)$*

If we take the real part of both sides of (3.15), we obtain

$$Re G(\mathbf{p}, E) = \mathscr{P}\int_0^\infty d\epsilon \left\{\frac{\rho^+(\mathbf{p}, \epsilon)}{E-(\mu+\epsilon)} + \frac{\rho^-(\mathbf{p}, \epsilon)}{E-(\mu-\epsilon)}\right\}.$$

We can let both integrals extend to $-\infty$ since the numerators vanish for negative ϵ. In the first integral, let $\epsilon+\mu = E'$; in the second integral let $\mu-\epsilon = E'$. Then

$$ReG(\mathbf{p}, E) = \mathscr{P}\int_{-\infty}^{\infty} dE' \frac{\rho^{+}(\mathbf{p}, E'-\mu)+\rho^{-}(\mathbf{p}, \mu-E')}{E-E'}.$$

Observe that either $\rho^{+}(\mathbf{p}, E'-\mu)$ or $\rho^{-}(\mathbf{p}, \mu-E')$ is zero for every E'. The numerator can now be written in terms of $ImG(\mathbf{p}, E)$ from (3.17) giving

$$ReG(\mathbf{p}, E) = \frac{1}{\pi}\mathscr{P}\left(\int_{\mu}^{\infty} dE' - \int_{-\infty}^{\mu} dE'\right) \cdot \frac{ImG(\mathbf{p}, E')}{E'-E}, \quad (3.19)$$

a dispersion relation relating the real and imaginary parts of $G(\mathbf{p}, E)$.* So far let me emphasize that we have assumed nothing about our system other than that $\mu^{N}-\mu^{N-1} = O(1/N)$.

4. *Analytic Properties of G(p)*

From the spectral representation, it is possible to say something about the analytic properties of $G(\mathbf{p}, E)$ as a function of the complex variable E for fixed $\mathbf{p}$. In (3.15), let $\epsilon \to -\epsilon$ in the ρ^{-} term. Then

$$G(\mathbf{p}, E) = \int_{-\infty}^{0} \frac{\rho^{-}(\mathbf{p}, -\epsilon)}{\epsilon-(E-\mu)+i\eta} d\epsilon - \int_{0}^{\infty} \frac{\rho^{+}(\mathbf{p}, \epsilon)}{\epsilon-(E-\mu)-i\eta} d\epsilon$$

* This is a more awkward dispersion relation than the usual kind which would be of the form

$$ReG(\mathbf{p}, E) = \frac{1}{\pi}\mathscr{P}\int_{-\infty}^{\infty} dE' \frac{ImG(\mathbf{p}, E')}{E'-E} \quad (3.19')$$

and which is given incorrectly by Galitskii and Migdal. The dispersion relation (3.19′) holds for a Green function defined in terms of the retarded commutator,

$$G(\mathbf{p}, E) = i\int_{-\infty}^{0} dt \exp(+iEt)\langle\Psi_0|[C_p(t), C_p^{\dagger}(0)]|\Psi_0\rangle,$$

which has somewhat simpler analytic properties than those derived from a T-product but which is not so simply related to a perturbation theory formalism. I shall discuss a retarded commutator's analytic properties in connection with the polarization propagator somewhat later.

which defines $G(\mathbf{p}, E)$ for real E. We may combine these two integrals by letting $E' = \epsilon + \mu$ and introducing a function $F(\mathbf{p}, E')$:

$$\begin{aligned} F(\mathbf{p}, E') &= +\rho^{-}(\mathbf{p}, \mu - E'), \qquad E' < \mu \\ &= -\rho^{+}(\mathbf{p}, E' - \mu), \qquad E' > \mu. \end{aligned} \tag{3.20}$$

Then, for E real, (3.20) is equivalent to

$$G(\mathbf{p}, E) = \int_{C_t} \frac{F(\mathbf{p}, E')}{E' - E} dE', \tag{3.21}$$

where the contour on the complex plane is as indicated.*

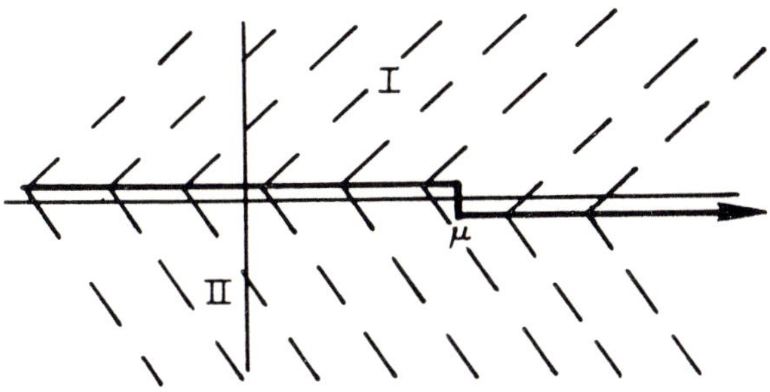

I should like to make several remarks:

(a) This contour divides the complex plane into two regions marked I and II. The integral on the right in (3.21) defines two distinct analytic functions of E, $f_{\mathrm{I}}(\mathbf{p}, E)$ and $f_{\mathrm{II}}(\mathbf{p}, E)$, one being analytic for all E in region I and the other analytic for all E in region II.†

* The retarded Green function defined in (3.19′) has the entire contour lying just below the real axis and so is analytic on the real axis and in the whole upper half plane.

† There is a well-known theorem in the theory of analytic functions (proved for example in Knopp, *Theory of Functions*, Dover Publications, Inc. N.Y. Vol. I, pp. 62 *ff*.) that an integral of the form $\int_C \varphi(\zeta)(\zeta - z)^{-1} d\zeta$ defines an analytic function for every point z not on C, providing $\varphi(\zeta)$ is a function defined and continuous along C. Although Knopp's proof requires that C be finite, it is readily extendable to infinite C if $\int_C \varphi(\zeta) d\zeta$ converges, which it does here.

(*b*) Furthermore, when E is real and $> \mu$, it lies in I so that $G(\mathbf{p}, E) = f_{\text{I}}(\mathbf{p}, E)$; when E is real and $< \mu$, it lies in II so that $G(\mathbf{p}, E) = f_{\text{II}}(\mathbf{p}, E)$.

(c) The function

$$\int_{C_t} \frac{F(\mathbf{p}, E')dE'}{E'-E}$$

goes into its complex conjugate as E crosses the contour. We'll show this for $E > \mu$. Then

$$\int_{C_t} \frac{F(\mathbf{p}, E')dE'}{E'-E} = \int_{-\infty}^{\infty} \frac{F(\mathbf{p}, E')}{E'-E-i\eta} dE'$$

$$= \mathscr{P} \int_{-\infty}^{\infty} \frac{F(\mathbf{p}, E')}{E'-E} + \pi i F(\mathbf{p}, E). \quad (3.22)$$

Now suppose $E \to E - 2i\eta$, i.e., E crosses the contour. We have

$$\int_{C_t} \frac{F(\mathbf{p}, E')dE'}{E'-(E-2i\eta)} = \int_{-\infty}^{\infty} \frac{F(\mathbf{p}, E')}{E'-E+i\eta} dE'$$

$$= \mathscr{P} \int_{-\infty}^{\infty} \frac{F(\mathbf{p}, E')}{E'-E} dE' - \pi i F(\mathbf{p}, E). \quad (3.22')$$

Comparison of (3.22) and (3.22′) proves the assertion.*

(d) Now for $E_0 > \mu$, $G(\mathbf{p}, E_0 - 2i\eta) = f_{\text{II}}(\mathbf{p}, E_0 - 2i\eta)$, i.e., $G(\mathbf{p}, E_0 - 2i\eta)$ is the analytic continuation of $G(\mathbf{p}, E < \mu)$,

* A slightly different way of formulating the analytic properties of $G(p)$ is as follows:

$$G(\mathbf{p}, E) = \int_{-\infty}^{\infty} d\epsilon \frac{F(\mathbf{p}, \epsilon)}{\epsilon - E}$$

defining $G(\mathbf{p}, E)$ for E both above and below the real axis. The real axis is a branch line. $G(\mathbf{p}, E)$ for E real is defined by

$$G(\mathbf{p}, E > \mu) = \lim_{\eta \to 0^+} G(\mathbf{p}, E + i\eta),$$

$$G(\mathbf{p}, E < \mu) = \lim_{\eta \to 0^+} G(\mathbf{p}, E - i\eta).$$

Remark (c) then follows by inspection.

to a point just below the contour at E_0. In words, (3.22) means:

$G^*(p, E > \mu)$ and $G(\mathbf{p}, E < \mu)$ are the same analytic function $f_{\mathrm{II}}(\mathbf{p}, E)$. Similarly $G(\mathbf{p}, E > \mu)$ and $G^*(\mathbf{p}, E < \mu)$ are the same analytic function $f_{\mathrm{I}}(\mathbf{p}, E)$.

6. *Meaning of* $G(\mathbf{p}, t)$, *the Asymptotic Behavior for Large t*

Suppose we let t become very large and positive. What can we say about $G(\mathbf{p}, t)$ and $G(\mathbf{p}, E)$?

Without Interactions: We have seen first that $G(\mathbf{p}, t)$ is the probability amplitude that, if there is a particle with momentum $\mathbf{p}$, $|\mathbf{p}| > p_F$ at $t = 0$, then there is a particle with momentum $\mathbf{p}$ at time t. We have also seen that $\rho^+(\mathbf{p}, \epsilon)$ is a delta function in ϵ at $\epsilon = \epsilon(\mathbf{p}) - \mu$, on the real axis. We can approximate this delta function with an expression of the form

$$\delta[\epsilon - \{\epsilon(\mathbf{p}) - \mu\}] \simeq \frac{1}{\pi} \frac{\Gamma}{[\epsilon - \{\epsilon(\mathbf{p}) - \mu\}]^2 + \Gamma^2} = \frac{1}{2\pi}\left\{\frac{i}{\epsilon - \{\epsilon(\mathbf{p}) - \mu\} + i\Gamma} - \frac{i}{\epsilon - \{\epsilon(\mathbf{p}) - \mu\} - i\Gamma}\right\}$$

in which Γ is understood to tend to zero.

With Interactions: Converting the sum in (3.12a) into an integral with the aid of (3.13a), we have

$$G(\mathbf{p}, t) = -i\, e^{-i\mu t} \int_0^\infty d\epsilon\, \rho^+(\mathbf{p}, \epsilon)\, e^{-i\epsilon t}, \text{ for } t > 0. \quad (3.23)$$

Now with interactions we may expect that $\rho^+(\mathbf{p}, \epsilon)$ is strongly peaked on the real axis near $\epsilon = \epsilon(\mathbf{p}) - \mu$. Suppose the nearest simple pole to the real axis is at $\bar{\epsilon}(\mathbf{p}) - \mu - i\Gamma_p$ with a residue $A_p/2\pi$. Then, because $\rho^+(\mathbf{p}, \epsilon)$ is real for ϵ on the real axis, there must be another simple pole at $\bar{\epsilon}(\mathbf{p}) - \mu + i\Gamma_p$ with residue $A_p^*/2\pi$. In other words, we may write $\rho^+(\mathbf{p}, \epsilon)$ as

$$\rho^+(\mathbf{p}, \epsilon) \simeq \frac{A_p/2\pi}{\epsilon - [\bar{\epsilon}(\mathbf{p}) - \mu - i\Gamma_p]} + \frac{A_p^*/2\pi}{\epsilon - [\bar{\epsilon}(\mathbf{p}) - \mu + i\Gamma_p]} + \ldots, \quad (3.24)$$

where the dots represent contributions from other singularities further from the real axis. A comparison with the noninteracting case suggests that, for weak interactions, Γ_p will be small compared with

$\bar{\epsilon}(\mathbf{p}) - \mu$ and the quantities A_p and $A_p{}^*$ will be almost purely imaginary having values i and $-i$. It is just when $\rho^+(\mathbf{p}, \epsilon)$ has this structure that a simple interpretation for $G(\mathbf{p}, t)$ is possible. Assuming this, let us now carry out the energy integration along the positive real axis indicated in (3.23). Rationalizing,

$$\rho^+(\mathbf{p}, \epsilon) \simeq \frac{1}{\pi} \frac{[\epsilon - \{\bar{\epsilon}(\mathbf{p}) - \mu\}] Re A_p + \Gamma_p Im\, A_p}{[\epsilon - \{\bar{\epsilon}(\mathbf{p}) - \mu\}]^2 + \Gamma_p{}^2},$$

$$G(\mathbf{p}, t) = -\, i\, e^{-i\mu t} \int_0^\infty d\epsilon\, \rho^+(\mathbf{p}, \epsilon)\, e^{-i\epsilon t}$$

$$= -\, i\, e^{-i\mu t} \left(\oint_{\circlearrowright} + \int_{\downarrow} \right) d\epsilon\, \rho^+(\mathbf{p}, \epsilon)\, e^{-i\epsilon t}.$$

The integral on the closed contour can be done by residues giving

$$G(\mathbf{p}, t) = -\, e^{-i\bar{\epsilon}(\mathbf{p})t} e^{-\Gamma_p t}\, A_p - \frac{i\, e^{-i\mu t}}{\pi} \times$$

$$\left\{ Re A_p \int_0^{-i\infty} d\epsilon \frac{[\epsilon - (\bar{\epsilon} - \mu)] e^{-i\epsilon t}}{[\epsilon - (\bar{\epsilon} - \mu)]^2 + \Gamma_p{}^2} + Im\, A_p \Gamma_p \int_0^{-i\infty} d\epsilon \frac{e^{-i\epsilon t}}{[\epsilon - (\bar{\epsilon} - \mu)]^2 + \Gamma_p{}^2} \right\}.$$

The first integral on the right is of the form

$$\int_0^{-i\infty} d\epsilon \frac{(\epsilon - \omega)\, e^{-i\epsilon t}}{(\epsilon - \omega)^2 + \Gamma^2} \simeq -\, i \int_0^\infty d\lambda \frac{e^{-\lambda t}}{(i\lambda + \omega)},$$

since Γ^2 can be neglected compared with $(\epsilon - \omega)^2$,

$$= i \int_0^\infty d\lambda \frac{e^{-\lambda t}}{\omega} \left(1 - \frac{i\lambda}{\omega} + \ldots \right)$$

yielding an asymptotic series

$$= \frac{i}{\omega t} - \frac{i}{(\omega t)^2} + \ldots = \frac{i}{(\bar{\epsilon} - \mu) t} - \ldots$$

The second integral on the right is the form

$$\int_0^{-i\infty} d\epsilon \frac{\Gamma e^{-i\epsilon t}}{(\epsilon - \omega)^2 + \Gamma^2} \simeq i \int_0^\infty d\lambda \frac{e^{-\lambda t} \Gamma}{(i\lambda + \omega)^2} = \frac{i}{\omega^2} \frac{\Gamma}{t} + \ldots$$

$$\simeq \frac{i\Gamma}{(\bar{\epsilon} - \mu)} \frac{1}{(\bar{\epsilon} - \mu) t} + \ldots$$

Then

$$G(\mathbf{p}, t) = - e^{-i\bar{\epsilon}(p)t} e^{-\Gamma_p t} A_p + \frac{e^{-i\mu t}}{\pi}\left[\left(Re A_p + Im A_p \frac{\Gamma_p}{\bar{\epsilon}-\mu}\right)\frac{1}{(\bar{\epsilon}-\mu)t} + \ldots\right].$$

For times so large that $(\bar{\epsilon}-\mu)t \gg 1$ but not so large that $\exp(-\Gamma_p t) \simeq 1/(\bar{\epsilon}-\mu)t$, we see that only the first term contributes to $G(\mathbf{p}, t)$. We thus have the important result that the pole of $\rho^+(\mathbf{p}, \epsilon)$ in the lower half plane nearest the real axis gives the energy and the damping of the elementary excitation of momentum $\mathbf{p}$ for $|\mathbf{p}| > p_F$. In the same way one can see that the poles of $\rho^-(\mathbf{p}, \epsilon)$ in the upper half plane give the energies and damping of elementary excitations for $|\mathbf{p}| < p_F$.

It is possible to express this conclusion in terms of $G(\mathbf{p}, E)$ itself. Recall (3.18):

$$\begin{aligned}\rho^+(\mathbf{p}, \epsilon) &= -\frac{1}{\pi} Im\, G(\mathbf{p}, \epsilon+\mu) \qquad \epsilon > 0,\\ &= -\frac{1}{\pi} Im\, G(\mathbf{p}, E) \qquad E > \mu,\\ &= -\frac{1}{2\pi i}[G(\mathbf{p}, E) - G^*(\mathbf{p}, E)].\end{aligned}$$

But for $E > \mu$, $G^*(\mathbf{p}, E)$ is analytic in the lower half plane $(=f_{II})$. *Thus the elementary excitations for $|\mathbf{p}| > p_F$ have energies and damping given by the poles of the analytic continuation of $G(\mathbf{p}, E)$ (for $E > \mu$) into the lower half-plane. Similarly, the elementary excitations for $|\mathbf{p}| < p_F$ have energies and damping given by the poles of the analytic continuation of $G(\mathbf{p}, E)$ (for $E < \mu$) into the upper half-plane.*

E. The Adiabatic Theorem, the S-Matrix, and Green Functions in the Interaction Picture

1. *The Interaction Picture and the Adiabatic Theorem*

To calculate the Green functions, we should like to know the exact ground state Ψ_0. In this and the following sections, I shall concern myself with the perturbation theoretic method for constructing Ψ_0. In this respect, the discussion wil be clearly limited, and will not *directly* include the important cases of superconductivity and liquid helium. However, this approach is systematic and gives a lot of physical insight, so it is the one I shall present in the limited time offered by these lectures. As I've mentioned, a lot of this formalism *can* be applied to the problems of liquid helium and superconductivity, but only after certain modifications.

The method for constructing Ψ_0 will be to start at $t = -\infty$ with our system in the ground state Φ_0 of the noninteracting part, H_0, of the Hamiltonian, to switch on the interaction H_1 extremely slowly, and to find the solution $\Psi_0(t)$ of Schrödinger's equation with the initial condition $\Psi_0(-\infty) = \Phi_0$. The adiabatic theorem then states that, under certain conditions (discussed below), $\Psi_0 = \Psi_0(0)$ is the ground state of the full Hamiltonian H. Just what part of H we include in H_0 and what part we include in H_1 is more or less arbitrary. We may, for example add a one-particle potential to the kinetic energy, calling all this H_0, and subtract it from the sum of the external and two-body potentials, calling all this H_1. Or, in the problem without external potential, we may just call the kinetic energy H_0 and the potential energy H_1.

Thus we think of the Hamiltonian

$$H_t = H_0 + f(t)H_1$$

in which the function $f(t) \to 0$ as $t \to \pm\infty$ and increases very slowly to the value unity as $t \to 0$. For example, $f(t) = \exp(-\eta|t|)$ is a popular choice where η is a positive infinitesimal.

In the Schrödinger picture, the equation for $\Psi_0{}^S(t)$ is

$$i\frac{\partial\Psi_0{}^S(t)}{\partial t} = \{H_0 + f(t)H_1\}\Psi_0{}^S(t). \tag{3.26}$$

(a) *Interaction Picture.* It is convenient, in solving (3.26), to go to the "interaction picture" (it is often called the interaction representation) to get rid of the H_0 term. Let us then define a state vector $\Psi_0{}^I(t)$ by

$$\Psi_0{}^I(t) = e^{iH_0t}\,\Psi_0{}^S(t) \quad \text{or} \quad \Psi_0{}^S(t) = e^{-iH_0t}\,\Psi_0{}^I(t). \tag{3.27}$$

In the absence of interactions, $\Psi_0{}^I$ is time-independent. Substituting (3.27) in (3.26), we have

$$i\frac{\partial\Psi_0{}^I(t)}{\partial t} = f(t)H_1(t)\Psi_0{}^I(t), \tag{3.28}$$

where

$$H_1(t) = e^{iH_0t} H_1 e^{-iH_0t}. \tag{3.28'}$$

These are the fundamental equations of the interaction picture corresponding to (2.1) and (2.1′) and to (2.3) and 2.3′) in the Schrödinger and Heisenberg pictures.

(b) *Converting to an Integral Equation.* It is more convenient to work with an operator that takes us from $\Psi_0{}^{\mathrm{I}}(t')$ to $\Psi_0{}^{\mathrm{I}}(t)$, than to work directly with the state vector $\Psi_0{}^{\mathrm{I}}(t)$. Thus define $U(t, t'; f)$ by the equation

$$\Psi_0{}^{\mathrm{I}}(t) = U(t, t'; f)\Psi_0{}^{\mathrm{I}}(t'). \tag{3.29}$$

Substituting this definition in (3.28) we obtain a differential equation for $U(t, t'; f)$:

$$i\frac{\partial}{\partial t}U(t, t'; f) = f(t)H_1(t)U(t, t'; f) \tag{3.30a}$$

with the boundary condition

$$U(t, t; f) = 1. \tag{3.30b}$$

If we integrate both sides of (3.30a) from t' to t, we have the integral equation

$$U(t, t'; f) = 1 - i\int_{t'}^{t} dt'' f(t'')H_1(t'')U(t'', t'; f), \tag{3.31}$$

which incorporates the condition (3.30b). From now on, let's assume that $f(t) = \exp(-\eta|t|)$, $\eta \to 0+$, and omit explicit reference to f. The combination $f(t)H_1(t)$ we'll call $\mathscr{H}_1(t)$.

(c) *Some Properties of* $U(t, t'; f)$. I shall assert a series of properties of the U operator without proof. The proofs are not hard, and the properties are certainly not surprising.

(1) $U^{-1}(t, t') = U(t', t)$, i.e., the inverse of going from t' to t is going from t to t'.

(2) U is unitary, i.e., $UU^\dagger = 1$. (This is true at least when there is no state like a Bardeen–Cooper–Schrieffer state splitting off.)

(3) $U(t, t_2) = U(t, t_1)U(t_1, t_2)$.

(4) $U(0, -\infty)\Phi_0 \equiv \Psi_0$ defines a solution of $H\Psi_0 = (\text{const})\Psi_0$. This is the adiabatic theorem and really is rigorous only if the unperturbed state Φ_0 lies a finite energy below the excited

states (not the case in a metal). Even if the theorem is valid, it may be that the state Ψ_0 is not the ground state of the interacting system. A simple example of this is a free electron gas (with no two-body interactions) at absolute zero when the "interaction" is the interaction of each electron's spin with an external magnetic field, $-\boldsymbol{\sigma}\cdot\mathbf{H}$. Then $\Psi_0 = \Phi_0$ (since the perturbation has no off-diagonal matrix elements), but Ψ_0 is not the state of lowest energy in the presence of a magnetic field.* For a proof of the adiabatic theorem, see Gell-Mann and Low.[3]

(5) $\Psi_0 = U(0, \infty)U(\infty, -\infty)\,\Phi_0$, which follows from (3) and (4). We call $U(\infty, -\infty) \equiv S$ the "S-matrix".

(6) $S\Phi_0 = \exp(i\phi)\Phi_0$ or $\langle\Phi_0|S|\Phi_0\rangle = \exp(i\phi)$. After turning the interaction on and then off, we must return within a factor, to the same state (if all is done infinitely slowly and Φ_0 is a non-degenerate state). By unitarity, the factor must have unit modulus, so this is the most general form for $S\Phi_0$.

2. *The Iteration Solution for $U(t, t')$ and the S-Matrix in the Interaction Picture*

We can get a formal solution for U by iterating the integral equation (3.31). Thus in successive orders we find

$$U_0 = 1, \tag{3.32a}$$

$$U_1 = 1 - i\int_{t'}^{t} dt_1 \mathscr{H}_1(t_1), \tag{3.32b}$$

$$U_2 = 1 - i\int_{t'}^{t} dt_1 \mathscr{H}_1(t_1) + (-i)^2 \int_{t'}^{t} dt_1 \int_{t'}^{t_1} dt_2 \mathscr{H}_1(t_1)\mathscr{H}_1(t_2), \tag{3.32c}$$

etc. . . .

* This difficulty, encountered in treating the Pauli paramagnetism of an electron gas, may be overcome either by doing the perturbation theory at finite temperatures (see Chapter V) and then taking the limit $T \to 0$, or by making the external magnetic field vary sinusoidally and then letting the wavelength of the variation tend to infinity.

With the T ordering symbol, we can simplify integrals like the last one in U_2:

$$\int_{t'}^{t} dt_1 \int_{t'}^{t_1} dt_2 \ldots \int_{t'}^{t_{n-1}} dt_n \mathscr{H}_1(t_1) \ldots \mathscr{H}_1(t_n)$$

$$= \frac{1}{n!} \int_{t'}^{t} dt_1 \int_{t'}^{t} dt_2 \ldots \int_{t'}^{t} dt_n T(\mathscr{H}_1(t_1) \ldots \mathscr{H}_1(t_n)).$$

The $1/n!$ occurs because there are $n!$ ways of ordering the times $t_1, \ldots, t_n$ all giving the same contribution to the integral on the right, but only one of these orderings is present in the integral on the left. Actually, the T-product not only orders operators in time, but also introduces a factor (-1) for every swap of neighboring ψ's and/or $\bar{\psi}$'s. These (-1)'s always occur an even number of times, however, when we reorder operators like $\mathscr{H}_1(t_1)$ and $\mathscr{H}_1(t_2)$, because the $\mathscr{H}$'s will always involve even numbers of ψ's and $\bar{\psi}$'s.

The infinite series to which the sequence (3.32) converges (*if* it converges) may be written very compactly as

$$U(t, t') = T\left(\exp -i \int_{t'}^{t} dt'' \mathscr{H}_1(t'')\right). \tag{3.33}$$

By this we mean to expand the exponential in a power series in its exponent, and then to time-order each term.

In this notation, the S-matrix is

$$S = T\left(\exp -i \int_{-\infty}^{\infty} dt \mathscr{H}_1(t)\right). \tag{3.34}$$

3. *The Iteration Solution for Green Functions in the Interaction Picture*

Just as it is simple to get a formal solution for the S-matrix in the interaction picture, it is also possible to get formal expressions for the various Green functions in this picture. Again the formal expressions are somewhat restricted in validity because of the iteration procedure used to construct them.

Recall the definition of the one-particle Green function,

$$G(\mathbf{r}t, \mathbf{r}'t') = -i\langle \Psi_0 | T(\psi(\mathbf{r}t)\bar{\psi}(\mathbf{r}'t')) | \Psi_0 \rangle, \tag{3.35}$$

in which ψ and $\bar{\psi}$ are Heisenberg operators.

First, recall our assertion that

$$\Psi_0 = U(0, -\infty)\Phi_0 \tag{3.36a}$$

is a stationary state, and recall also that

$$\Psi_0 = U(0, \infty)U(\infty, -\infty)\Phi_0 = U(0, \infty)S\Phi_0. \tag{3.36b}$$

Also from property (6) of U we have

$$\langle S\Phi_0| = e^{-i\phi} \langle\Phi_0| = \frac{\langle\Phi_0|}{\langle\Phi_0|S|\Phi_0\rangle}. \tag{3.36c}$$

Substituting these relations into (3.35), one gets

$$G(\mathbf{r}t, \mathbf{r}'t') = -i\langle S\Phi_0|U^\dagger(0, \infty)T(\psi(\mathbf{r}t)\bar{\psi}(\mathbf{r}'t'))U(0, -\infty)|\Phi_0\rangle$$

or

$$G(\mathbf{r}t, \mathbf{r}'t') = -i\frac{\langle\Phi_0|U(\infty, 0)T(\psi(\mathbf{r}t)\bar{\psi}(\mathbf{r}'t'))U(0, -\infty)|\Phi_0\rangle}{\langle\Phi_0|S|\Phi_0\rangle}. \tag{3.37}$$

Now let us write ψ and $\bar{\psi}$ in terms of interaction picture operators. To do this recall that for any state vectors Ψ_a and Ψ_b

$$\langle\Psi_a{}^{\mathrm{I}}(t)|\ \mathcal{O}^{\mathrm{I}}(t)|\Psi_b{}^{\mathrm{I}}(t)\rangle = \langle\Psi_a{}^H|\ \mathcal{O}^H(t)|\Psi_b{}^H\rangle, \tag{3.38)*$$

and recall, too, that at $t = 0$

$$\Psi_a{}^S = \Psi_a{}^{\mathrm{I}} = \Psi_a{}^H, \text{ etc., and } \mathcal{O}^S = \mathcal{O}^{\mathrm{I}} = \mathcal{O}^H. \tag{3.39}$$

Using the U operator we may write

$$\langle\Psi_a{}^{\mathrm{I}}(t)|\ \mathcal{O}^{\mathrm{I}}(t)|\Psi_b{}^{\mathrm{I}}(t)\rangle = \langle\Psi_a{}^{\mathrm{I}}(0)|U(0, t)\mathcal{O}^{\mathrm{I}}(t)U(t, 0)|\Psi_b{}^{\mathrm{I}}(0)\rangle. \tag{3.40}$$

Comparing this with (3.38) we have

$$U(0, t)\ \mathcal{O}^{\mathrm{I}}(t)U(t, 0) = \mathcal{O}^H(t). \tag{3.41}$$

Finally, we can rewrite (3.37):

$$G(x, x') =$$

$$-i\ \langle\Phi_0|U(\infty, 0)\ T\ \{(U(0, t)\psi^{\mathrm{I}}(\mathbf{r}t)U(t, 0))\ (U(0, t')\bar{\psi}^{\mathrm{I}}(\mathbf{r}'t')U(t', 0))$$

$$U(0, -\infty)\}|\Phi_0\rangle/\langle\Phi_0|S|\Phi_0\rangle.$$

* I haven't proved this but it follows immediately from the definition.

If $t > t'$, reading from the right, we first go from $-\infty$ to t', let $\bar{\psi}$ act, then go from t' to t, let ψ act, then go from t to ∞. If $t < t'$, we go from $-\infty$ to t, let ψ act, go from t to t', let $\bar{\psi}$ act, and finally go to ∞. We may combine both cases in the expression

$$G(x, x') = \frac{-i\langle\Phi_0|T(\psi^{\mathrm{I}}(\mathbf{r}t)\bar{\psi}^{\mathrm{I}}(\mathbf{r}'t')S)|\Phi_0\rangle}{\langle\Phi_0|S|\Phi_0\rangle}. \tag{3.42}$$

This is the desired expression for $G(x, x')$ in the interaction picture. In this expression, it is understood that S depends on all values of the time as in (3.34) and the time-ordering operator breaks it up into parts occurring before, between, and after ψ and $\bar{\psi}$.

F. Diagrammatic Analysis of the Iteration Solution

1. *Creation and Annihilation Operators and Normal Products*

Let us take the entire potential energy part of H (except for the infinite self-energy $-\frac{1}{2}Nv(\mathbf{r} = 0)$, which we include in H_0) as the perturbation H_1. That is

$$H_1 = \int \bar{\psi}(x)w(\mathbf{r})\psi(x)d^3r + \tfrac{1}{2}\int\int \bar{\psi}(x)\psi(x)v(x-x')\bar{\psi}(x')\psi(x')d^4x'\, d^3r. \tag{3.43a}$$

We shall assume, though it is not strictly necessary, that

$$v(x-x') = v(\mathbf{r}-\mathbf{r}')\delta(t-t'). \tag{3.43b}$$

This is certainly the case for the instantaneous Coulomb interaction. $w(\mathbf{r})$ could be a periodic potential or a set of impurity scattering centers, or an external field.

The ground state of the "unperturbed" Hamiltonian, H_0, is the free Fermi sea, Φ_0. The operators $\bar{\psi}$ and ψ are creation and annihilation operators with reference to the bare vacuum. Let us now define new creation and annihilation operators with respect to the Fermi sea or "Fermi vacuum".

To see these operators explicitly, it is convenient to expand in the operators $C_p(t)$ and $C_p^\dagger(t)$:

$$\psi(\mathbf{r}t) = \underbrace{\Omega^{-\frac{1}{2}} \sum_{p<p_F} e^{i\mathbf{p}\cdot\mathbf{r}}\, C_p(t)}_{\equiv\, \psi^+,\ \text{creates holes}} + \underbrace{\Omega^{-\frac{1}{2}} \sum_{p>p_F} e^{i\mathbf{p}\cdot\mathbf{r}}\, C_p(t)}_{\equiv\, \psi^-,\ \text{destroys electrons}}, \qquad (3.44\text{a})$$

$$\bar{\psi}(\mathbf{r}t) = \underbrace{\Omega^{-\frac{1}{2}} \sum_{p<p_F} e^{-i\mathbf{p}\cdot\mathbf{r}}\, C_p^\dagger(t)}_{\equiv\, \bar{\psi}^-,\ \text{destroys holes}} + \underbrace{\Omega^{-\frac{1}{2}} \sum_{p>p_F} e^{-i\mathbf{p}\cdot\mathbf{r}}\, C_p^\dagger(t)}_{\equiv\, \bar{\psi}^+,\ \text{creates electrons}}. \qquad (3.44\text{b})$$

Thus ψ^- and $\bar{\psi}^-$ are the new annihilation operators and ψ^+ and $\bar{\psi}^+$ are the new creation operators.

It is important to notice that

$$\overline{\psi^+} = \bar{\psi}^-, \quad \text{not} \quad \bar{\psi}^+,$$

and

$$\overline{\psi^-} = \bar{\psi}^+, \quad \text{not} \quad \bar{\psi}^-.$$

The "particles" created by $\bar{\psi}^+$ and ψ^+ are electrons and holes. They are *not* conserved. This has the advantage, however, that we may often deal with but a few particles. This decomposition into creation and annihilation operators is not changed with time, in the interaction or Schrödinger pictures. In these pictures, if C_p destroys an electron above the Fermi surface at $t = 0$, for example, so does $C_p(t)$, etc.

We now want to introduce the idea of a "Normal Product". When we go to evaluate $\langle\Phi_0|S|\Phi_0\rangle$, we expand S and get products of the form

$$\bar{\psi}(x_1)\psi(x_1)\bar{\psi}(x_2)\psi(x_2)\ldots$$

(There will also be other factors, integrations, etc.) Similarly in evaluating the expression $\langle\Phi_0|T(\bar{\psi}(x)\psi(x')S)|\Phi_0\rangle$ we get the same kind of products with the operators $\bar{\psi}(x)$ and $\psi(x')$ inserted somewhere in among the other operators. It would be particularly convenient in evaluating the (Fermi) vacuum expectation of such products of operators if all the annihilation operators stood to the right of all the creation operators, for then the vacuum expectation would be zero. Thus we use the following plan:

(a) Express the ψ's and $\bar{\psi}$'s in terms of ψ^-, ψ^+, $\bar{\psi}^-$, $\bar{\psi}^+$.
(b) Anticommute the ψ^-'s and $\bar{\psi}^-$'s to the right, the ψ^+'s and $\bar{\psi}^+$'s to the left, giving several anticommutators in the process.
(c) Take the vacuum expectation value of the result.
(d) The only terms which will *not* vanish will be those in which no operators are remaining, only anti commutators (which, it turns out, are not operators, but c-numbers).

To be more precise, we define a *Normal Product* of a set of operators:

$N(\bar{\psi}(x_1)\psi(x_2)\ldots) \equiv$ product (or sum of products) of these operators in which
(a) ψ^+, $\bar{\psi}^+$ are moved to the left (anticommutators neglected),
(b) ψ^-, $\bar{\psi}^-$ are moved to the right (anticommutators neglected), and
(c) all sign changes on anticommuting are included.

Then

$$\langle\Phi_0|N(\bar{\psi}(x_1)\psi(x_2)\ldots)|\Phi_0\rangle = 0. \tag{3.45}$$

As an illustrative example, we have

$$\begin{aligned} N(\bar{\psi}(x)\psi(x')) &= N\{(\bar{\psi}^+(x)+\bar{\psi}^-(x))(\psi^+(x')+\psi^-(x'))\} \\ &= \bar{\psi}^+(x)\psi^+(x')+\bar{\psi}^+(x)\psi^-(x')+\bar{\psi}^-(x)\psi^-(x')-\psi^+(x')\bar{\psi}^-(x). \end{aligned}$$

Let's now define a *Contraction of a Pair of Operators*. This is the anticommutator we omit when we formally reorder a T-product of a pair of operators to get an N-product:

$$\underbracket{AB} = T(AB)-N(AB) \tag{3.46}$$

defines the contraction of A and B. For the operators in which we shall be interested, the contractions are c-numbers. This provides, in conjunction with (3.45), an easy way to compute a contraction:

$$\underbracket{AB} = \langle\Phi_0|T(AB)|\Phi_0\rangle \tag{3.46'}$$

The important contractions in the interaction picture are readily calculated if we recall that $\psi_I(\mathbf{r}, t) = \exp[iH_0t]\psi_H(\mathbf{r}, 0)\exp[-iH_0t]$

$$\underbracket{\psi(x)\bar{\psi}(x')} = iG_0(x, x'), \tag{3.47a}$$

$$\underbracket{\psi(x)\bar{\psi}(x)} = -\rho_0(x), \tag{3.47b}$$

the average electron density at x in the unperturbed ground state, and

$$\underbracket{\psi(x)\psi(x')} = \underbracket{\bar{\psi}(x)\bar{\psi}(x')} = 0. \tag{3.47c}$$

2. *Wick's Theorem and its Diagrammatic Interpretation*

The key theorem which allows us to give a clear and simple prescription for writing down the vacuum expectation of any T-product is known as Wick's theorem. I am not going to prove this theorem, because proofs can be found in all the standard textbooks in quantum field theory. Perhaps one of the clearest proofs is to be found in Akhiezer and Berestetsky. The theorem as we shall use it states that for any product of Fermi operators $ABC \ldots XYZ$,

$$\begin{aligned}
T(ABC \ldots XYZ)& \\
&= N(ABC \ldots XYZ) \\
&\quad + N(\underbracket{AB}C \ldots XYZ) + \ldots + N(ABC \ldots X\underbracket{YZ}) \\
&\quad + N(\underbracket{AB}\,\underbracket{CD} \ldots XYZ) + \ldots + N(ABC \ldots \underbracket{WX}\,\underbracket{YZ}) \\
&\quad + \ldots \\
&\quad \cdot \\
&\quad \cdot \\
&\quad \cdot \\
&\quad + \underbracket{AB}\,\underbracket{CD} \ldots \underbracket{WX}\,\underbracket{YZ} + \text{all other pairings.}
\end{aligned} \tag{3.48}$$

In the second line on the right, in each term we bring two operators together (by anticommuting, neglecting the anticommutators), replace the pair by its contraction (a c-number), and take the N-product of the remaining operators. We do this with all possible pairs, i.e., $n(n-1)/2$ if there are n operators, and so we get this many terms, each term still containing an N-product of the n-2 remaining operators. In the third line on the right, we choose two pairs in all possible ways, replacing them by their contractions, and leaving in each term an N-product of n-4 operators. And so on. In the last line, we pair all the operators in all possible ways (or "pairings"), replacing all pairs by their contractions and leaving no operators.

To compute the vacuum expectation of a T-product, we use Wick's theorem and find that only the completely contracted terms [last line on the right of (3.48)] contribute anything; all the others have normal products, which have zero vacuum expectations. That is,

$$\langle\Phi_0|T(ABC\ldots XYZ)|\Phi_0\rangle = \sum_{\text{all pairings}} \text{(product of contractions of all pairs including minus signs from reordering).}$$

As a first example, consider some of the terms which occur in calculating the vacuum expectation of the S-matrix.

$$\langle\Phi_0|S|\Phi_0\rangle = 1 - i\left\langle \Phi_0 \left| \int dt_1\ \mathscr{H}_1(t_1) \right| \Phi_0 \right\rangle + \frac{(-i)^2}{2!}\left\langle \Phi_0 \left| T\left(\int dt_1 \int dt_2\ \mathscr{H}_1(t_1)\mathscr{H}_1(t_2)\right) \right| \Phi_0 \right\rangle + \ldots$$

$$= 1 - i\left(\int d^4x_1\ e^{-\eta|t_1|}\ w(x_1)\langle\Phi_0|T(\bar{\psi}(x_1)\psi(x_1))|\Phi_0\rangle + \tfrac{1}{2}\int d^4x_1 d^4x_1'\ e^{-\eta|t_1|}\ v(x_1 - x_1') \langle\Phi_0|T(\bar{\psi}(x_1)\psi(x_1)\bar{\psi}(x_1')\psi(x_1'))|\Phi_0\rangle \right) + \frac{(-i)^2}{2!}(\ldots) + \ldots$$

The contributions from each term in the expansion are easy to calculate and each one can be represented by a diagram. Let us work through all the contributions from the first order term in the expansion just to get the feel of things. Then we'll state the general rules for writing down the contributions corresponding to different diagrams. The one-body potential, in first order, contributes only one term,

$$-i\int d^4x_1\ e^{-\eta|t_1|}\ w(x_1)\langle\Phi_0|\underbrace{\bar{\psi}(x_1)\psi(x_1)}|\Phi_0\rangle = \int d^4x_1\ e^{-\eta|t_1|}(-iw(x_1))\rho_0(x_1), \tag{3.50}$$

where

$$\rho_0(x) = \langle\Phi_0|\bar{\psi}(x)\psi(x)|\Phi_0\rangle$$

is the number of particles per unit volume at x in the noninteracting ground state Φ_0. If in general we represent each contraction of a ψ and a $\bar{\psi}$ by a line starting at the argument of $\bar{\psi}$ and ending at the argument of ψ (recall that the only nonzero contractions always involve one ψ and one $\bar{\psi}$), and each factor $-iw(x)$ by a dotted line from

x to a shaded circle representing the external source of potential, then the particular term (3.50) is represented by the diagram

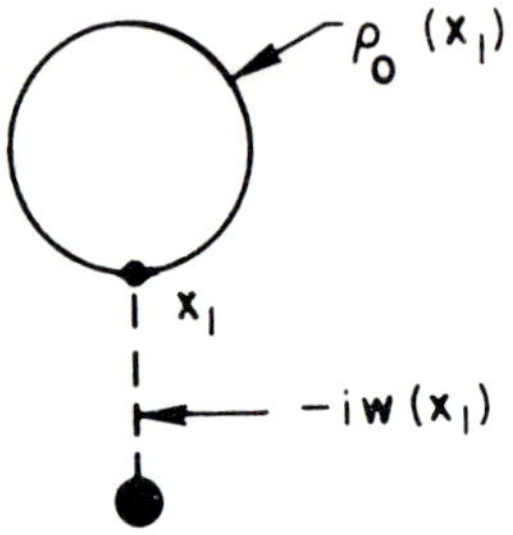

The two-body interaction, in first order, contributes two terms corresponding to the two possible pairings which give a nonzero contribution:

$$-\tfrac{1}{2}i\int d^4x_1 d^4x_1{}'\, e^{-\eta|t_1|}\, v(x_1-x_1{}')\{\langle\Phi_0|\underbrace{\bar\psi(x_1)\psi(x_1)}\;\underbrace{\bar\psi(x_1{}')\psi(x_1{}')}|\Phi_0\rangle$$

$$+\langle\Phi_0|\bar\psi(x_1)\psi(x_1)\bar\psi(x_1{}')\psi(x_1{}')|\Phi_0\rangle\} = -\tfrac{1}{2}\,i\int d^4x_1\, d^4x_1{}' e^{-\eta|t_1|}\, v(x_1-x_1{}')$$

$$\{\rho_0(x_1)\rho_0(x_1{}') + (-iG_0(x_1{}', x_1))\,(+iG_0(x_1, x_1{}'))\}. \qquad (3.51)$$

These two terms are represented by the following two diagrams:*

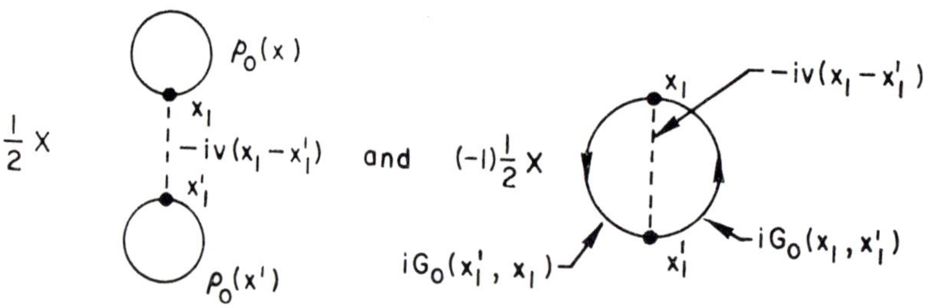

Here, the solid line from $x_1{}'$ to x_1 represents the contraction $\underbrace{\psi(x_1)\bar\psi(x_1{}')}$ and so has the factor $iG_0(x_1, x_1{}')$ associated with it. The arrow tells

* For a discussion of a certain complication that arises in the second of these diagrams, see subsection 8(d) of this section.

us that the line leaves x_1 and goes to x_1 i.e., the $\bar{\psi}$ operator acts at x_1' and the ψ operator acts at x_1. The extra factor (-1) arises from the need to anticommute $\psi(x_1')$ through three operators to get each ψ to the left of the $\bar{\psi}$ with which it is contracted.

All diagrams representing contributions to $\langle\Phi_0|S|\Phi_0\rangle$, such as those just drawn, are called vacuum fluctuation diagrams.

Now let us look at the contributions to

$$\langle\Phi_0|T(\psi(x)\bar{\psi}(x')S)|\Phi_0\rangle$$

in first order. The term

$$-i\int d^4x_1\, e^{-\eta|t_1|}\, w(x_1)\langle\Phi_0|T(\psi(x)\bar{\psi}(x')\bar{\psi}(x_1)\psi(x_1))|\Phi_0\rangle$$

has nonzero contributions from two different pairings:

$$\langle\Phi_0|\underbrace{\psi(x)\bar{\psi}(x')}\underbrace{\bar{\psi}(x_1)\psi(x_1)}|\Phi_0\rangle$$

corresponding to

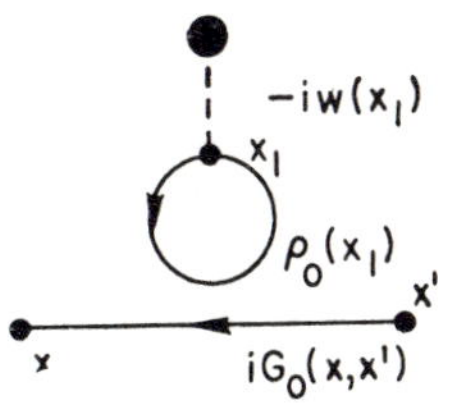

and

$$\langle\Phi_0|\psi(x)\bar{\psi}(x')\bar{\psi}(x_1)\psi(x_1)|\Phi_0\rangle$$

(with $\psi(x)$ contracted with $\bar{\psi}(x_1)$, and $\bar{\psi}(x')$ contracted with $\psi(x_1)$)

corresponding to

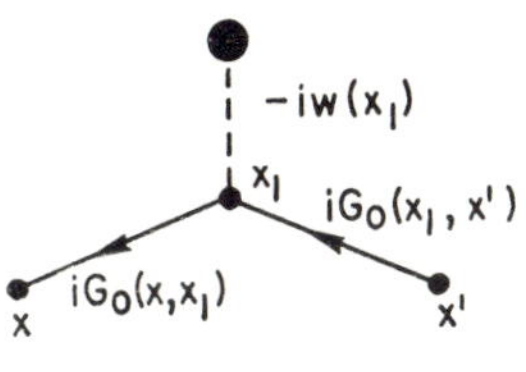

The term

$$-\tfrac{1}{2}i\int d^4x_1 d^4x_1' v(x_1-x_1')\, e^{-\eta|t_1|}$$
$$\langle\Phi_0|T(\psi(x)\bar{\psi}(x')\bar{\psi}(x_1)\psi(x_1)\bar{\psi}(x_1')\psi(x_1'))|\Phi_0\rangle$$

has nonzero contributions from six different pairings corresponding to the following six diagrams (we omit the factors associated with the various lines of the diagram):

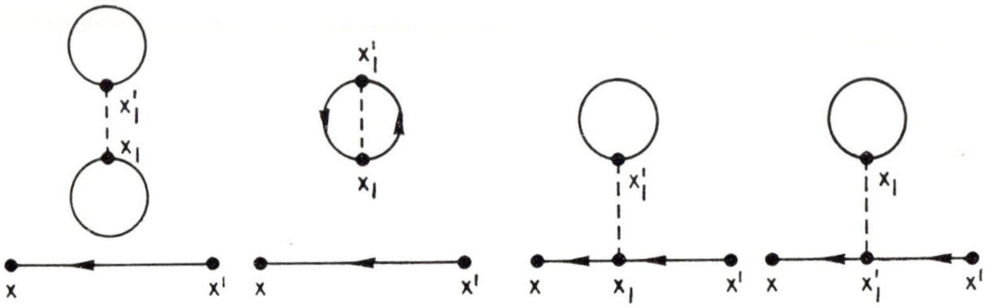

(a) (b) (c) (d)

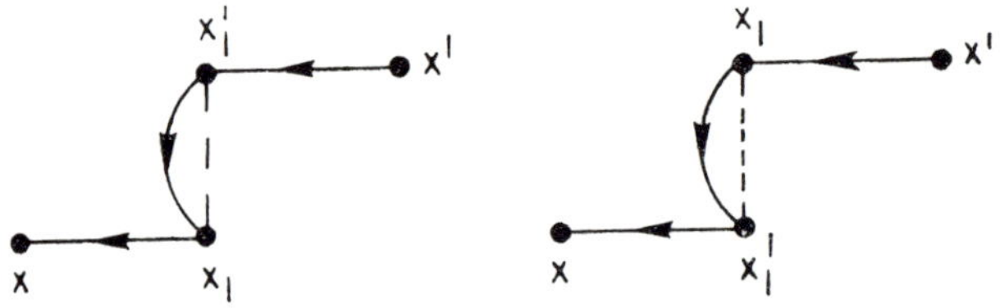

(e) (f)

An instructive exercise would be (1) to show that these are the only pairings with a nonzero contribution and, (2) to find the contribution from each pairing. For example, the contribution from the pairing represented by diagram (b) is

$$-\tfrac{1}{2}\int d^4x_1\, d^4x_1'\; e^{-\eta|t_1|} \left[\{iG_0(x, x')\}\{-iv(x_1-x_1')\}\{iG_0(x_1', x_1)\}\{iG_0(x_1, x_1')\}\right].$$

We see, then, for these examples (and it is true in general) that any term in the expansions of the numerator or denominator of (3.42) with any pairing can be represented by a diagram, and that the contribution of this term can be simply calculated from the diagram by associating factors like iG_0, ρ_0, $-iw$, or $-iv$ with various elements of the diagram. It is also true that any diagram constructed according to certain rules has associated with it some term in (3.42) with some pairing. This suggests the following procedure:

(a) Draw all diagrams possible according to certain rules.
(b) Write down the contribution from each diagram.
(c) Sum these contributions in both the numerator and denominator of the expression (3.42) for G.
(d) Divide the denominator explicitly into the numerator.

Let us discuss each of these points separately.

3. *How to Draw a Diagram in x-Space*

(a) For the contributions to the numerator of the r-particle Green function draw r points labelling them $x_1' \ldots x_r'$ (at which lines will start) and r points labelling them $x_1, x_2 \ldots x_r$ (at which lines will end), anywhere on the paper. Every diagram will ultimately have a directed solid line (a line with an arrow on it) leaving each point labelled with an x' and a directed solid line arriving at each point labelled with an x. For the vacuum diagrams, there are no points labelled with x's or x''s.

(b) To get a contribution to either numerator or denominator in which the one-body potential acts n_1 times, draw n_1 points labelling them $z_1, z_2 \ldots z_{n_1}$ and from each have a dotted line running to a small filled-in circle nearby (representing the source of the one-body potential).

(*c*) To get a contribution to either numerator or denominator in which the two-body potential acts n_2 times, draw n_2 pairs of points labelling them $y_1, y_1', y_2, y_2' \ldots y_{n_2}, y_{n_2}'$ and have a dotted line connecting each y with the corresponding y'.

(d) Add directed solid lines between points so that each "internal point" (i.e., a y, y' or z) has one solid line into and one solid line out of it, and each "external point" (i.e., an x or x') has one line out of it (if it is an x') or into it (if it is an x).

4. *Computing the Contribution of a Diagram in x-Space*

Once a diagram is drawn, it is very easy, in principle, to write down the contribution of that term in the expansion corresponding to that diagram. Here are the rules:

(*a*) For every element 

, write $iG_0(y_i - y_j)$.

(b) For every element 

, write $-iw(z_i)$.

(c) For every element  , write $-iv(y_i - y_i')$.*

(d) For every element  , write $\rho_0(y_i)$.

(e) Include a factor (-1) for every closed loop of solid lines in the diagram. The diagrams  and  have one and two closed loops respectively. The element  is not considered a closed loop. These minus signs come from an odd number of anticommutations needed to bring contracted operators together. The second term in (3.51) provides an example of a (-1) factor.

(f) Integrate over all values from $-\infty$ to ∞ of the four coordinates of every *internal* point.

5. *To Sum the Diagrams—The Counting Problem*

In principle we should like to draw all possible diagrams, computing the contribution from each, and sum these. By this I mean that, for a given n_1 and n_2, we label the ends of all the dotted lines and all the exterior points once and for all and then draw in the solid lines in all allowable ways.

First, remark that the positions of the points on the paper are arbitrary. Thus any two diagrams correspond to the same pairing if they are *topologically the same*, i.e., if one can be brought into the other simply by moving the points on the paper. For example, the following two diagrams are really the same.

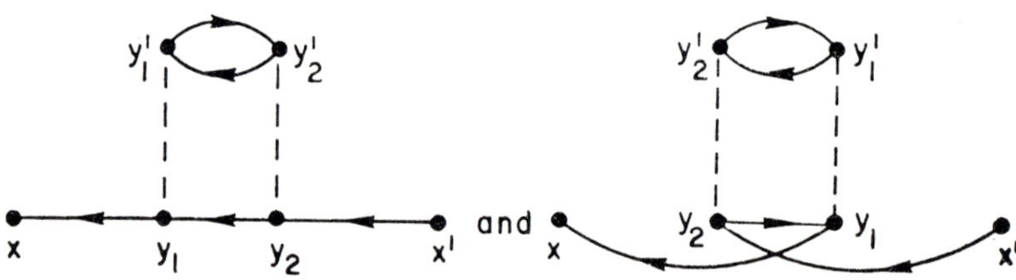

(a) (b)

* $v(y_i)$ is, in general, even in y, so such lines need not be directed, although it is usually convenient if they are.

Second observe that often two diagrams having their lines drawn in different ways differ only in having the labels on the ends of one or more two-body interaction lines swapped ($y_i \rightleftarrows y_i'$; $y_j \rightleftarrows y_j'$; etc.) and/or the labels on the ends of the one-body interaction lines permuted (z_i, z_j, z_k, etc. $\rightleftarrows z_{Pi}$, z_{Pj}, z_{Pk}, etc.) and/or the labels on the ends of the two-body interaction lines permuted (keeping y_i on the same dotted line as y_i', of course). Thus diagram (c), which appears to be different from (b) in the way solid lines are drawn, actually differs from (a) and hence from (b) (which is identical to (a)) only through a permutation of labels.

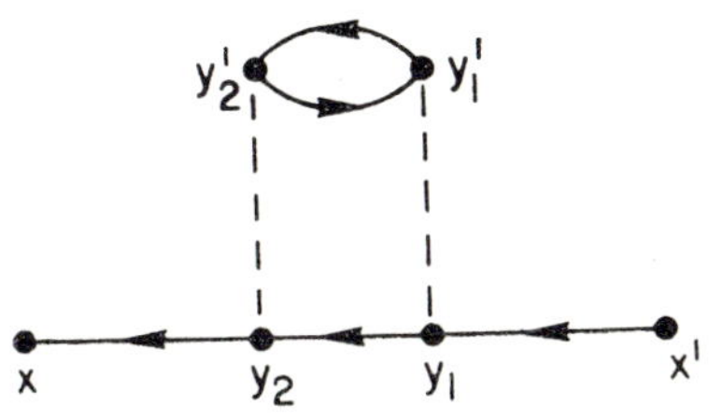

(c)

In such cases, let us say that the two diagrams have the *same structure* (i.e., they are of the *same structure* if, when we omit the labels of *internal* points, but not the arrows, the two diagrams are topologically the same). It is clear that all diagrams having the same structure make the same contribution because they differ only in the labelling of the internal points, and these labels are just dummy variables of integration. Thus, instead of calculating the contribution of every diagram and summing, we need only calculate the contribution of each structure, determine the number of diagrams having that structure, and add each contribution that many times.

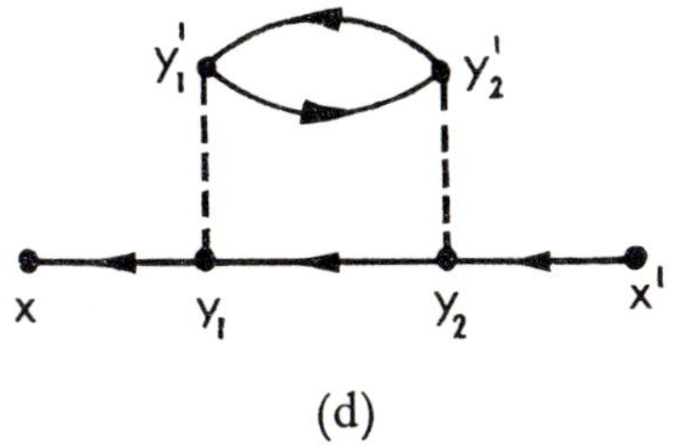

(d)

The existence of closed loops of solid lines in a diagram can be a source of confusion about whether or not two diagrams have the same structure. For example, diagram (d) is identical with (a) because the relative position of the two directed lines between y_1 and y_2' is completely arbitrary. On the other hand, the following two diagrams are different because one cannot be deformed into the other.

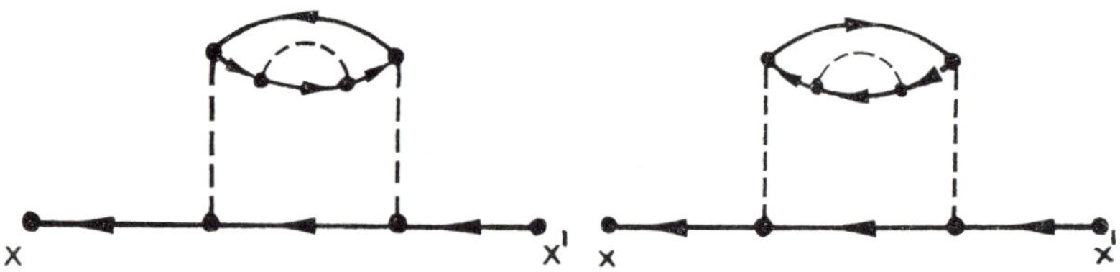

The following two diagrams are identical, however, because one can be deformed into the other by reflecting through the obvious vertical axis.

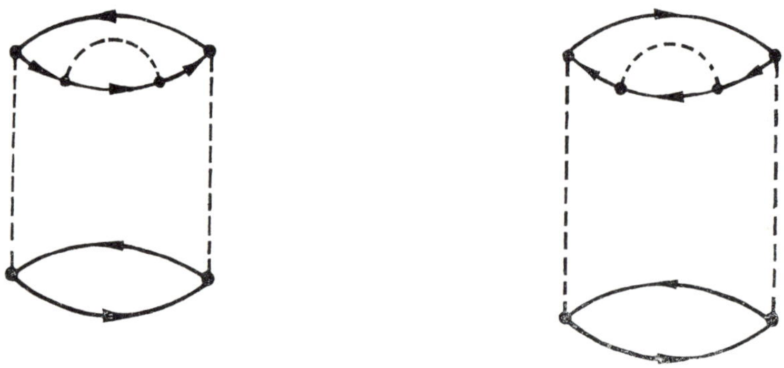

Let us return to the basic problem, which is to count the number of diagrams of a given structure. At first sight, we might guess that, for a structure with n_1 one-body lines and n_2 two-body lines, there are $2^{n_2}n_1!\,n_2!$ different diagrams corresponding to the two possibilities of arranging the labels on the ends of each of the n_2 two-body lines, the $n_1!$ permutations of the one-body lines and the $n_2!$ permutations of the two-body lines. Recall that $v(y_i-y_i')$ always is accompanied by a factor 1/2 giving a factor $(1/2)^{n_2}$ in the contribution of such a diagram. Also, in nth order, the expansion of the exponential in

S gives a factor $1/n!$ Finally in the nth order, there are $n!/n_1!n_2!$ ways in which we can get n_1 one-body interactions and n_2 two-body interactions. Combining all these factors we have

$$(\tfrac{1}{2})^{n_2}\frac{1}{n!}\,\frac{n!}{n_1!\;n_2!}$$

suggesting that we get all contributions if we compute the contribution of each structure and just add them all together.

Actually, this will give too high a contribution from many structures, because sometimes the swapping or permuting of labels may yield a diagram which is topologically the same as the original diagram. For example, the following two diagrams obtained by swapping labels on both dotted lines are really the same diagram. In

other words, our estimate of $2^{n_2}n_1!n_2!$ different diagrams from the same structure may be too high. Let us suppose we start with one diagram of a definite structure, call it A, and start permuting and swapping in all possible ways, counting the number of times (including the original diagram) that we obtain diagrams topologically the same as A. Let this number be called h. Now consider a diagram of this structure which is topologically distinct from A—call it B. Then there will also be h diagrams (including B) identical with B, since each of the labelling changes which left A unchanged will also leave B unchanged and so on. All in all there will be $2^{n_2}n_2!n_1!/h$ sets of identical diagrams, each set being distinct from all the others. And so the *contribution of a given structure should be multiplied by a factor* $1/h$ *before the contributions of all structures are added together.*

It might be worthwhile to rephrase this in the language of group theory for those who prefer a more precise mathematical statement of the situation. Let $R_1 \ldots R_{n_2}$ be the operations that swap the labels on the 1st, 2nd, ... n_2th two-body lines; let $P_1 \ldots P_{n_2!-1}$ be the operations that permute the two-body lines among themselves; let $Q_1 \ldots Q_{n_1!-1}$ be the operations that permute the one-body lines among themselves; and let E be the identity operation which neither permutes nor swaps any labels. Then the group defined as the direct product

$$\mathscr{G} \equiv (E, R_1) \times (E, R_2) \times \ldots \times (E, R_{n_2}) \times (E, P_1 \ldots P_{n_2!-1}) \times (E, Q_1 \ldots Q_{n_1!-1})$$

is a group of order $2^{n_2} n_2! \, n_1!$, the group of all allowable changes of the labels in a diagram of n_1 one-body lines and n_2 two-body lines. Consider all the elements of $\mathscr{G}$ which leave a standard diagram topologically unchanged. These elements form a normal subgroup, $\mathscr{H}$, whose order we call h. It is a normal subgroup because all the elements are homomorphic with the identity topological transformation. In fact, the factor group $\mathscr{G}/\mathscr{H}$, whose elements are the cosets of $\mathscr{H}$, is isomorphic with the group of distinct topological transformations. The number of distinct diagrams is the number of distinct topological transformations which is the order of the factor group $= 2^{n_2} n_2! n_1!/h$. To find h (if that is ever necessary) one can rely on the methods of group theory. It should be remarked, however, that, although the structure of G is defined once n_1 and n_2 are given, the subgroup $\mathscr{H}$ depends in addition on the particular structure of the diagram (for a given n_1 and n_2 there are usually several different structures with different h's).

Some examples may help to clarify this counting problem. Consider the following diagram:

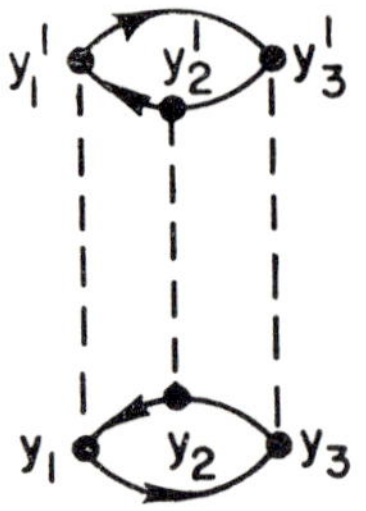

Let R_1, R_2, and R_3 swap the labels at the ends of the first, second, and third interaction lines respectively; let E, P_1, and P_2 be the three cyclic permutations of the interaction lines; and let S_{12}, S_{23}, and S_{31} be the operations interchanging lines 1 and 2, 2 and 3, and 3 and 1 respectively. Then $\mathscr{G}$ has $2^3 \times 3!$ different elements. The normal subgroup $\mathscr{H}$ which leaves the diagram topologically unchanged consists of the elements E, P_1, P_2, $R_1R_2R_3$, $R_1R_2R_3P_1$, and $R_1R_2R_3P_2$. In other words, $h = 6$ and there are $2^3 \times 3!/6 = 8$ distinct diagrams having this structure.

Similarly consider the diagram

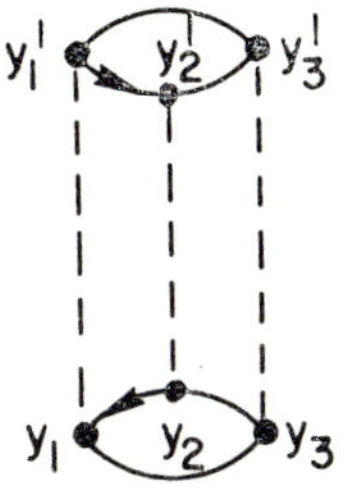

Now $\mathscr{H}$ consists of E, P_1, P_2, $R_1R_2R_3S_{12}$, $R_1R_2R_3S_{23}$ and $R_1R_2R_3S_{31}$, i.e., again $h = 6$ and there are eight distinct diagrams. Note, however, that these two examples are not only different diagrams, but different structures. One cannot obtain one from the other by any of the 48 allowable operations. Both structures must be considered in evaluating $\langle \Phi_0 | S | \Phi_0 \rangle$.

Still another example of a third-order diagram is

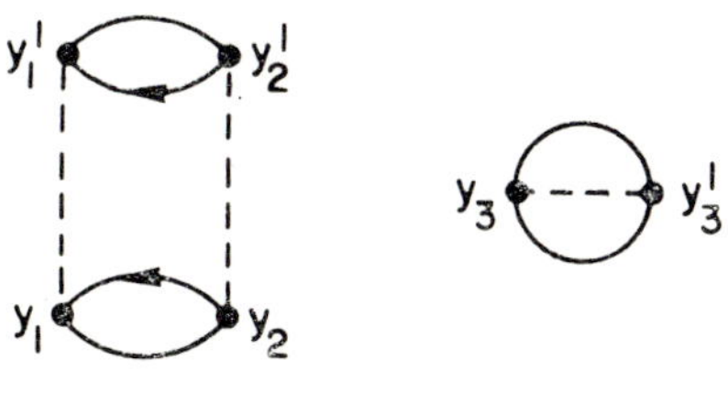

Now $\mathscr{H}$ consists of E, S_{12}, R_1R_2, $R_1R_2S_{12}$, R_3, $S_{12}R_3$, $R_1R_2R_3$, and $R_1R_2R_3S_{12}$ so that $h = 8$ and there are only six distinct diagrams. This example demonstrates that h depends on the particular structure as well as n_1 and n_2.

6. *Dividing Out the Vacuum Diagrams—the Linked Cluster Expansion*

We now have a prescription for writing down all the terms appearing in both the numerator (Green function diagrams) and the denominator (vacuum fluctuation diagrams) of (3.42). Let's now see that it is very easy to divide the denominator into the numerator. We shall make the simplifying (though not necessary) assumption that $w(\mathbf{r}) = 0$—i.e., we neglect the one-body potential.

First consider the denominator. It consists of a sum of all possible vacuum fluctuation structures, each with its appropriate weight factor $1/h$. For example, in first order we have

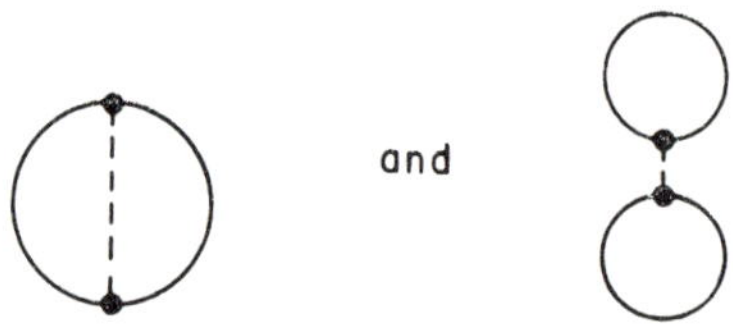

(1a) (1b)

and in second order we have

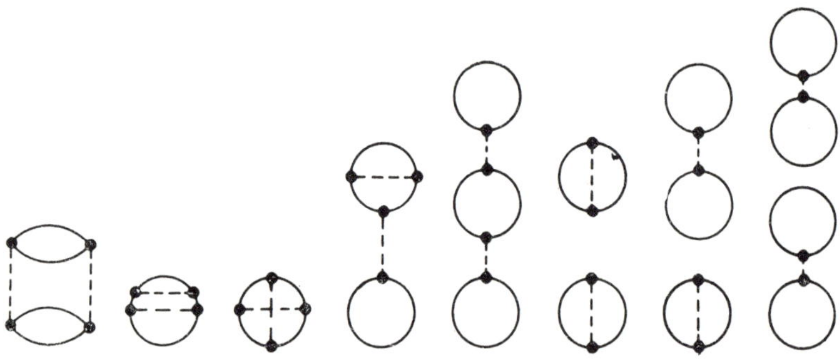

(2a) (2b) (2c) (2d) (2e) (2f) (2g) (2h)

Structures (2f)–(2h) differ from the others in that they each consist of two unconnected parts, but they must be included just like the rest.

Now consider the numerator for the one-particle Green function. Every diagram will at least consist of a part starting at x' and ending at x, a "proper Green function diagram". Some diagrams may also contain vacuum fluctuation parts not connected to the proper Green function part. For example, for the numerator we find in first order

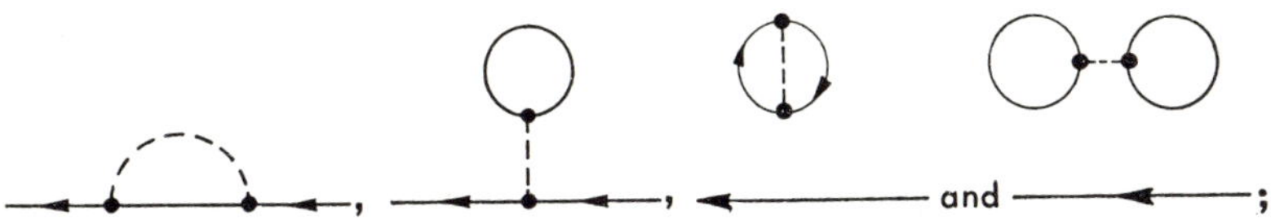

(1a) (1b) (1c) (1d)

and in second order we find

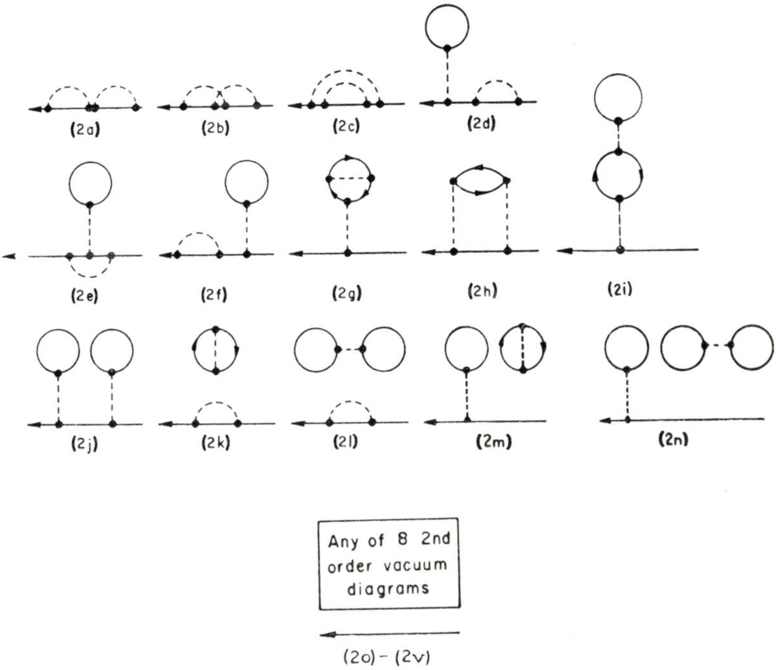

We see that diagrams (2k) to (2v) consist of two unconnected parts, one a vacuum fluctuation part. In fact, for every proper Green function diagram, there will be other diagrams obtained by adding to the proper Green function part all possible vacuum fluctuation diagrams.

We now wish to point out the important result that the contribution of any structure with two or more unconnected parts is the product of the contributions of all the separate parts, each considered as an isolated structure. It is obvious that the integral over all the internal vertices will factor into the integrals over the internal vertices of each part separately. It is perhaps somewhat less obvious that the weighting factor for the whole diagram is the product of the weighting factors for the separate parts. For simplicity, consider a diagram having a proper Green function part of order n and an unconnected vacuum fluctuation part of order n'. Such a diagram would enter in $n+n'$ order. What is the multiplicative factor in front of the integral for such a structure? First we get $1/(n+n')!$ from expanding the S-matrix and $(1/2)^{n+n'}$ because the two-body interaction acts $n+n'$ times (we are still neglecting the one-body force—to include it only complicates the argument a bit but does not invalidate the conclusion). These factors are just cancelled because from the permutation of the labels on the ends of the various interaction lines one obtains $(n+n')!$ and from swapping the labels on the ends of one or more lines one gets the factor $2^{n+n'}$, all just as if the diagram were connected. If we suppose there are h proper Green function subdiagrams which are topologically identical and h' vacuum fluctuation subdiagrams that are topologically identical, then it is clear that there are hh' whole diagrams that are topologically identical. Thus the contribution of the structure is multiplied by $(1/h)(1/h')$. In other words the contribution of an unconnected structure is the product of its various parts.

Since each proper Green function diagram can have with it any (or no) vacuum fluctuation diagram, we can sum over all vacuum fluctuation diagrams for a fixed proper Green function diagram, giving a factor $\langle\Phi_0|S|\Phi_0\rangle$ multiplying each proper Green function diagram. This factor is cancelled by the denominator of (3.42). (3.42) then consists of just the sum over all proper Green function diagrams, or "connected diagrams" or "linked clusters"; i.e.

$$G(x, x') = -i\langle\Phi_0|T(\psi(x)\bar{\psi}(x')S)\Phi_0\rangle_{\text{connected}},$$

or more generally

$$G(x_1 \dots x_n;\ x_1' \dots x_n') = (-i)^n \langle \Phi_0 | T(\psi(x_1) \dots \psi(x_n)\bar{\psi}(x_n') \dots \bar{\psi}(x_1')S)|\Phi_0\rangle_{\text{connected}} \tag{3.52}$$

and this is the linked cluster expansion for the Green functions.

7. *Diagrams and Rules in p-Space*

The calculation of diagrams and the summation of diagrams may be considerably simplified, if the unperturbed system is homogeneous and time-independent, by taking Fourier transforms of everything in sight. The result, as we shall now see, is to associate different quantities with the same diagrams, and to perform different and often simpler integrals. Let us first see how this works and then state the rules succinctly.

Introduce the following Fourier integral representations:

$$G_0(x, x') = G_0(x-x') = \int \frac{d^4p}{(2\pi)^4} e^{ip \cdot (x-x')} G_0(p),$$

$$v(x-x') = \int \frac{d^4q}{(2\pi)^4} e^{iq \cdot (x-x')} v(q),$$

$$w(x) = \int \frac{d^4q}{(2\pi)^4} e^{iq \cdot x} w(q).$$

Recall that $v(\mathbf{q} = 0) = 0$ because of cancellation against the positive background charge. Also $w(q) = w(\mathbf{q})\delta(\nu)$, if the external potential is time independent.

Then each line in a diagram (—← or - - - -) can be considered to have

(a) either $iG(p)$, $-iv(q)$ or $-iw(q)$ on it,

(b) a factor $\exp(-ip \cdot x_1)$ [or $\exp(-iq \cdot x_1)$] at the beginning of it, and

(c) a factor $\exp(ip \cdot x_2)/(2\pi)^4$ [or $\exp(iq \cdot x_2)/(2\pi)^4$] at the end of it, where x_1 and x_2 are the space-time coordinates of the beginning and end of the line. It is immaterial which end of a dotted line we take to be the "beginning" or "end", but once a choice has been made, arrows should be added to these lines. The sense of the arrows is thus arbitrary; they only insure that opposite ends of dotted lines are treated consistently.

Let us now integrate over all internal space-time points (including the factors $1/(2\pi)^4$ from the particle lines ending at the x_i's). This gives $\delta\left(p_{\text{in}} - p_{\text{out}} \mp q_{\substack{\text{out}\\ \text{in}}}\right)$ at each vertex.

Now many of the remaining p and q integrals can be done giving explicit four-momentum conservation at every vertex.

Still remaining in the diagram are

(a) Nontrivial integrals over some q's and some p's, and
(b) Integrations involving $\int d^4p' \exp(-ip \cdot x')$ and $(2\pi)^{-4} \int d^4p \exp(ip \cdot x)$ from the outgoing and incoming external lines.

All the diagram excluding the integrations (b) represents the contribution of this diagram to $G(p', p)$, the double transform of $G(x', x)$ (if total momentum is not conserved, which would be the case if there were an external potential). In case there is no external potential, then the total momentum into the diagram must equal the total momentum out, giving an additional $\delta(p'-p)$ for the diagram's contribution to the one-particle Green function. The p' integration can be done leaving from (b) above $(2\pi)^{-4} \int d^4p \exp\{ip \,.\, (x-x')\} \times$ (rest of diagram). Thus the rest of the diagram is the contribution to $G(p)$.

One is thus led to the following rules for constructing and computing diagrams in p-space:

(a) Draw all distinct structures, as in x-space, with the same factors $1/h$. All particle lines must be directed. Different ways of directing them, that are not topologically equivalent, give distinct structures. Direct interaction lines for convenience.
(b) Assign p's and q's to all lines so that the sum of the four-momenta entering a vertex equals the sum of the four-momenta leaving a vertex.
(c) For every element $\xleftarrow{p}$ include a factor $iG_0(p)$.
(d) For every element $-\overset{q}{\leftarrow}--\bullet$ or $--\overset{q}{\leftarrow}--$ include a factor $-iw(q)$ or $-iv(q)$ respectively.
(e) Include a factor (-1) for every closed loop of solid lines in the diagram (except diagram elements like $------\circlearrowleft$ which vanish anyway because $v(q = 0) = 0$).

(f) Compute

$$\int \frac{d^4q}{(2\pi)^4} \cdots \int \frac{d^4p}{(2\pi)^4} \cdots \times \text{ (all factors from } c, d, \text{ and } e),$$

the integral going over all internal momenta that are still free.

For example, the following diagram gives the following contribution to $iG(p)$.

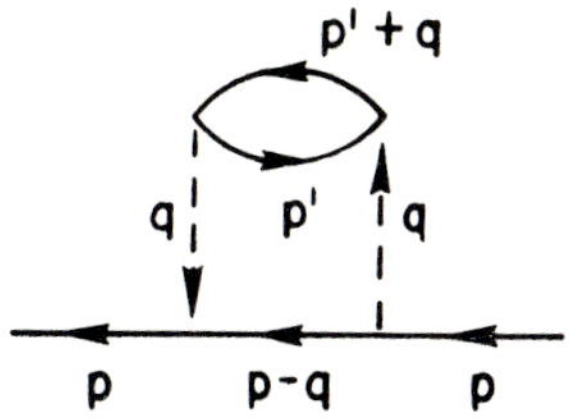

$$(-1)\int \frac{d^4q}{(2\pi)^4} \int \frac{d^4p'}{(2\pi)^4} \{iG_0(p)\}\{-iv(q)\}\{iG_0(p-q)\}\{iG_0(p')\}$$

$$\times \{iG_0(p'+q)\}\{-iv(q)\}\{iG_0(p)\}$$

$$= i\{G_0(p)\}^2 \int \frac{d^4q}{(2\pi)^4} \{v(q)\}^2 G_0(p-q) \quad \int \frac{d^4p'}{(2\pi)^4} G_0(p')G_0(p'+q).$$

We might emphasize that the fact we integrate over all momentum space for each free momentum does not allow us to omit structures which are topologically distinct because of different senses of directed lines. Thus, the following two structures must be included separately, their contributions being quite different:

8. *Random Remarks on Diagrams*

(a) *Deformability*. We have already remarked that it really makes no difference in computing the diagram just where the points are placed on the paper. There are at least two somewhat systematic ways which have certain advantages. One convention is to think of time (in x-space diagrams) running in a certain direction, say from right to left. Interaction lines, which we have assumed to be instantaneous, are then vertical. This is a useful convention for getting a physical interpretation of the process to which a given diagram corresponds. Another convention is to draw all the continuous particle lines as long straight lines and then the interaction lines will not all be straight and vertical. The physical interpretations become a bit less obvious, but it is easier to keep track of all the diagrams this way.

As an example, consider the second order contribution to G, assuming $w(x) = 0$:

Time running to left | Particle lines stretched out

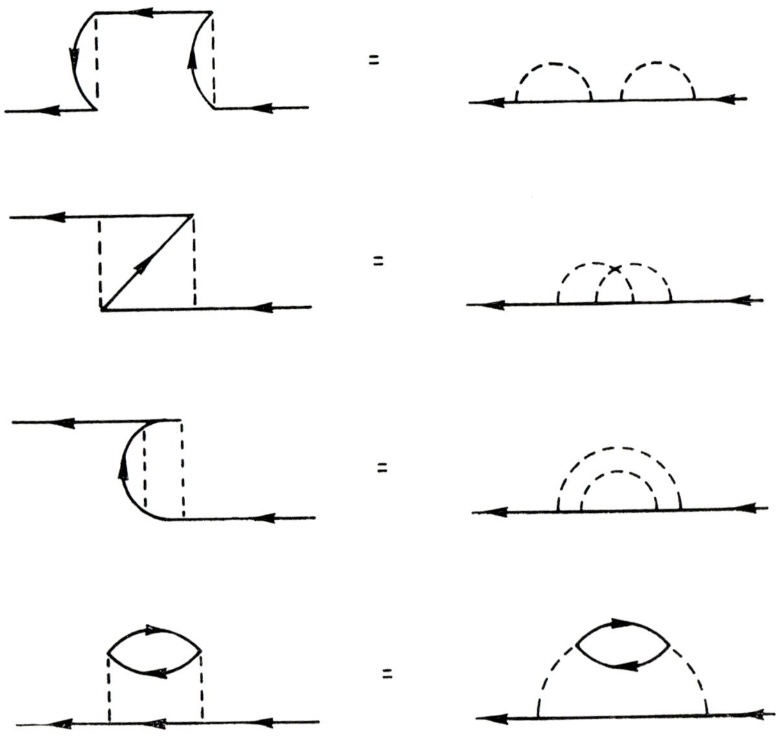

(b) *Interpretation of a Few Diagrams and Elements of Diagrams.* Here are a few parts of diagrams which occur very frequently:

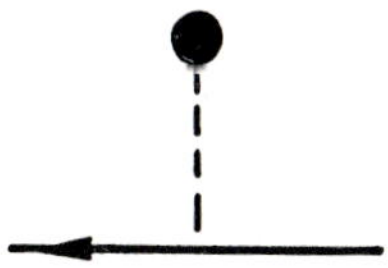

Particle scattered by external potential or background charge.

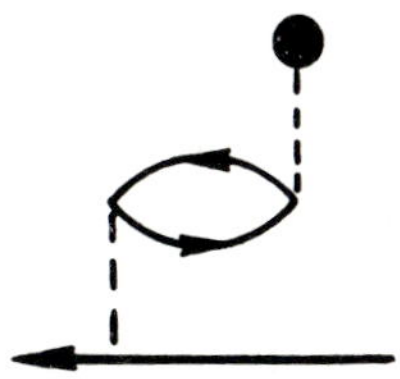

External potential excites an electron and hole which are then annihilated scattering an incident particle. This contributes to the screening of the external potential.

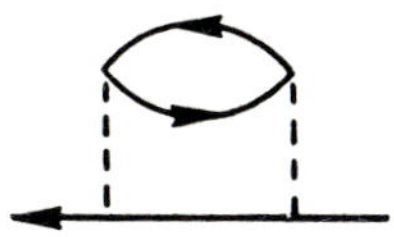

Particle excites an e–h pair, then de-excites it. This contributes to the self-energy of the particle.

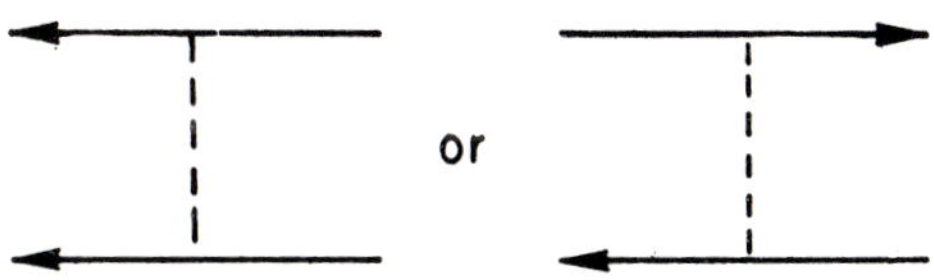

Two particles scatter.

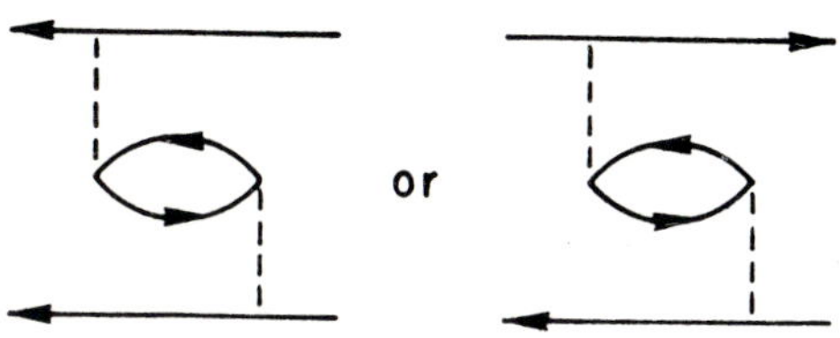

Two particles scatter with one exciting an e–h pair and the other de-exciting it. This contributes to screening the potential between particles.

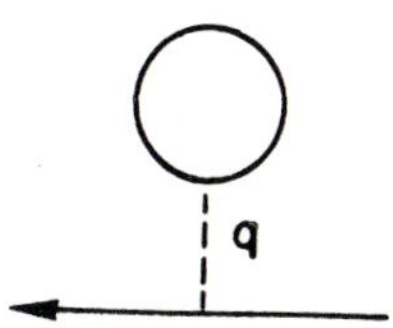

Particles scattered by Fermi sea. If 4–momentum is conserved, $q=0$. Also, if Fermi sea and background charge are assumed uniform $v(\mathbf{q}=0)=0$ and such diagrams give nothing.

(c) *The Exclusion Principle.* A careful reflection will show that we *seem* to have included diagrams which violate the Pauli principle. Since such diagrams should clearly be omitted, one might wonder if an error has been made, and if so where? Actually, whenever a diagram is included that violates the Pauli principle, another diagram is included that also violates the Pauli principle and that makes exactly the same contribution, except for an additional factor of (-1) due to one more or less closed loop. Here are two examples if we think of the diagrams as having been transformed to $(\mathbf{p}, t)$ space. Then for certain orderings of the time of the vertices, the exclusion principle is violated in the following pairs of diagrams which just cancel each other.

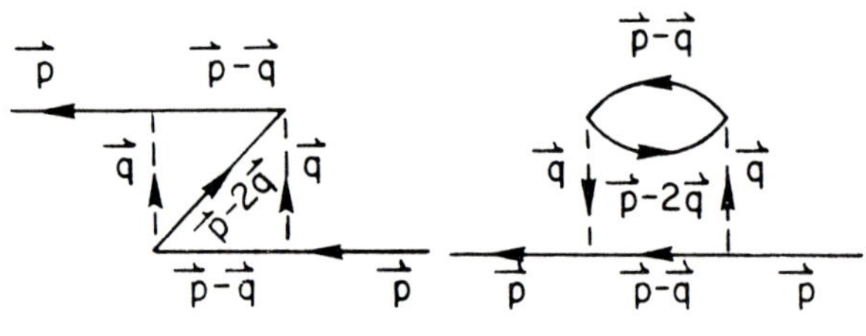

(two particles in $\mathbf{p}-\mathbf{q}$)

Also, the following two diagrams just cancel

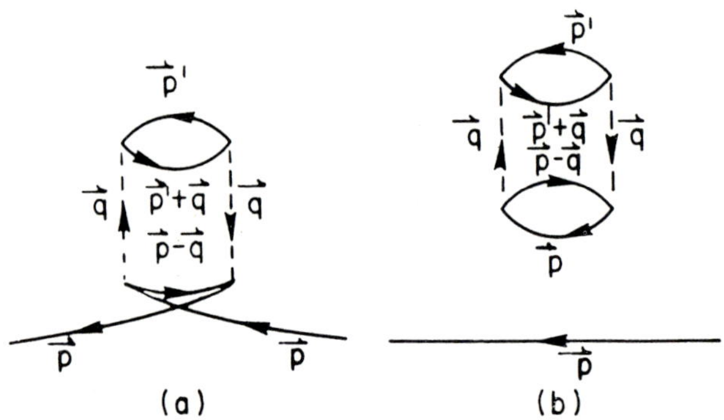

(two particles in $\mathbf{p}$)

The second example is interesting because it includes an unconnected diagram, (*b*). Neither part of the unconnected diagram violates the exclusion principle separately; in particular, the vacuum part will definitely appear in $\langle \Phi_0 | S | \Phi_0 \rangle$. To be able to divide it out of the numerator of (3.42), we must include the case shown above which violates the exclusion principle. For this reason we also include (a), which also violates the exclusion principle and just cancels (b).

(d) *Green's Function Between Two Simultaneous Points.* A special difficulty arises in computing the contribution of the diagram (or subdiagram)

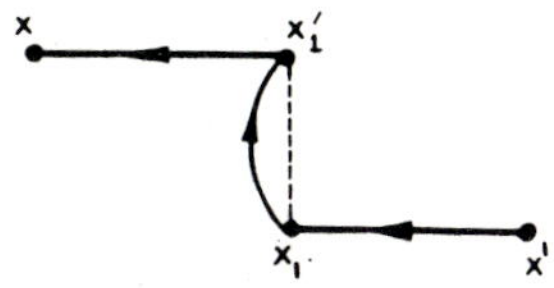

or

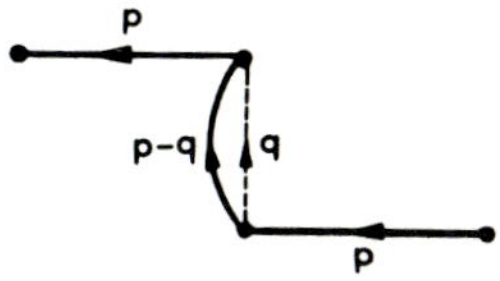

because, in x-space the two ends of the interaction line occur at the same time for instantaneous interactions, yet $iG_0(\mathbf{r}t, \mathbf{r}'t)$ isn't defined uniquely in this special case. Of course, in any diagram, the integration over intermediate times will include isolated points where the times at opposite ends of particle lines are equal, but in general the values of G_0 for these points don't affect the integral. However, in the diagram above, the instantaneous nature of the interaction makes the entire contribution come from such an isolated point. When such a structure arises from some order in the perturbation expansion, as it always does in the present discussion, it really represents one-half the contribution of two diagrams in which $t_1' = t_1^+$:

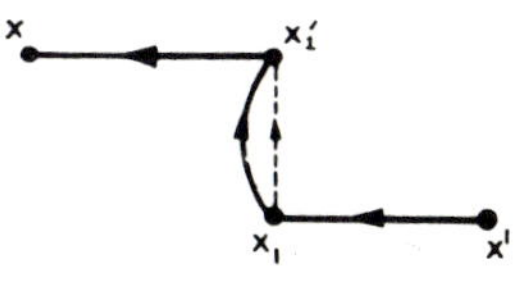

and

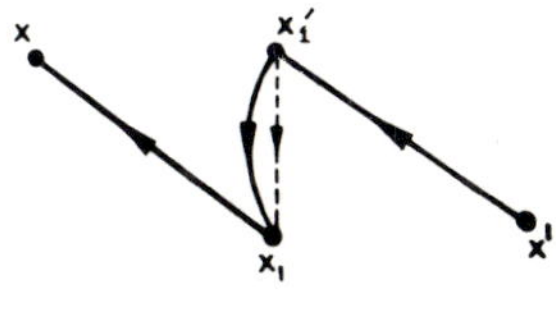

Thus by $G_0(\mathbf{r}t, \mathbf{r}'t)$ we must understand $\frac{1}{2}\{G_0(\mathbf{r}t, \mathbf{r}'t^+) + G_0(\mathbf{r}t, \mathbf{r}'t^-)\}$. In momentum space this means that instead of $G_0(p-q)$, we must understand $\frac{1}{2}G_0(p-q)\{\exp(i\nu\eta)+\exp(-i\nu\eta)\}$, where η is a positive infinitesimal. More explicitly, if one calculates the contribution of the subdiagram

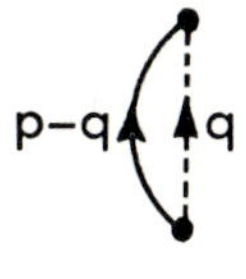

with this understanding, one finds

$$-\frac{i}{2}\int\frac{d^3q}{(2\pi)^3}v(\mathbf{q})[(1-f_{p-q})e^{i(E-\epsilon_{p-q})\eta}-f_{p-q}e^{-i(E-\epsilon_{p-q})\eta}],$$

where f_p is the Fermi function $= 1-2\delta_p$. For many purposes such as the calculation of the so-called single particle self-energy (see Section IIIG1), one may set $\eta = 0$. For other purposes, as in various methods of calculating the ground state energy (see sections IIIH 1 and 2), these diagrams arise in a slightly different way so that the factors involving η also appear differently. At this point it is sufficient if one is forewarned against this complication.

(e) *Anomalous Graphs of Kohn and Luttinger*. Consider the diagram

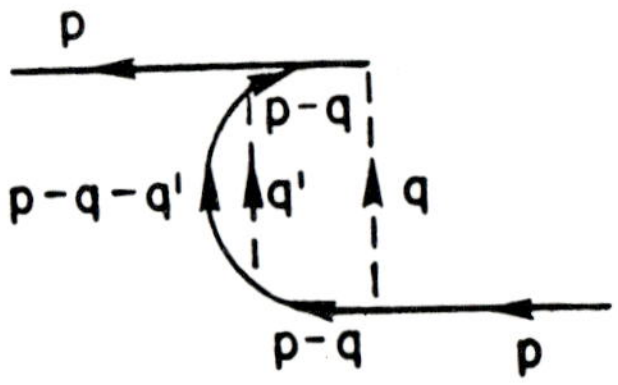

We might think its contribution would be zero because it requires both an electron and a hole to have the same momentum $p-q$. Kohn and Luttinger[4, 5] argue that this is not the case if such a graph is viewed as the limit, as Temp. $\to 0$, of a graph in a finite temperature theory. The crucial point, they find, is that the limit of infinite volume (with fixed particle density) should be taken *before* the limit $T \to 0$. Then this graph will in general give a finite contribution as $T \to 0$. However, a consistent treatment of finite T and the chemical potential $\mu(T)$ shows that for systems of spin 1/2 which are also isotropic (therefore *neither* systems with periodic potential *nor* with uniform external fields), we may ignore these anomalous diagrams. It is not known if these are the most general conditions.

G. INTEGRAL EQUATIONS FOR PARTIALLY SUMMING DIAGRAMS

At first sight, one might think that a systematic approximation procedure for calculating $G(p)$ would be to sum the contributions of all diagrams of the first few orders. It turns out, however, that the analytic properties (poles, branch points, asymptotic behavior)

which are the things of real significance, are *not* meaningfully affected if we include only a finite number of diagrams. The lowest approximation to get some interesting new physics, it turns out, must include an infinite number of diagrams. Let us look at three different kinds of infinite partial summations which can be effected, and which lead to interesting results. In each case, the partial summations lead to integral equations (or, if in a convenient representation, algebraic equations).

1. *Dyson's Equation*

Consider all the diagrams contributing to $G(p)$.*

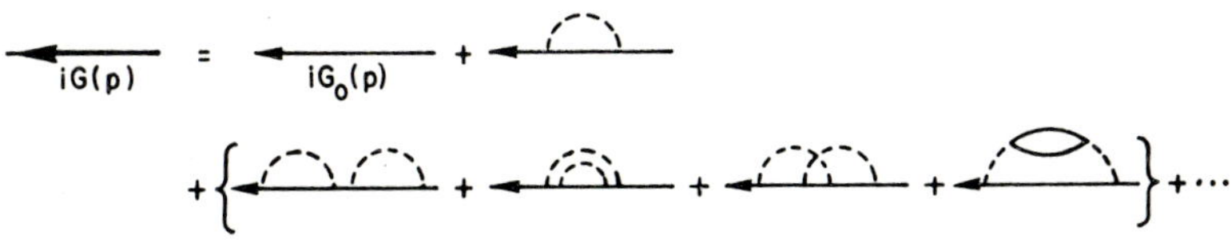

We shall derive an equation for $G(p)$ in terms of a quantity $\Sigma(p)$, which is a sum over a much simpler set of diagrams. Except for the zeroth order diagram, every diagram can be divided into three parts which we can represent schematically:

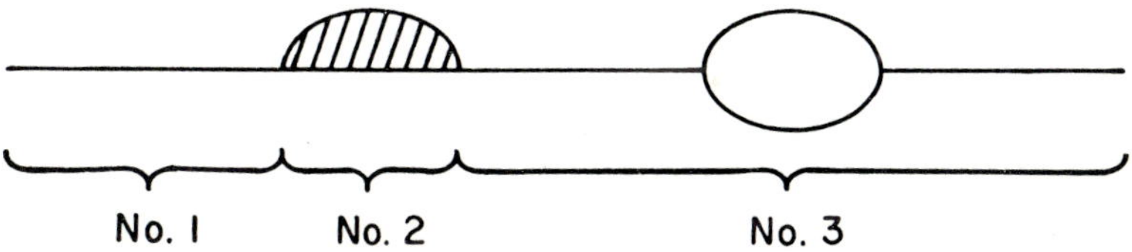

The shaded bump represents any subdiagram that is connected to an incoming and an outgoing line and that has the property that it cannot be divided into two parts by breaking only one solid line. The following subdiagrams are of this type:

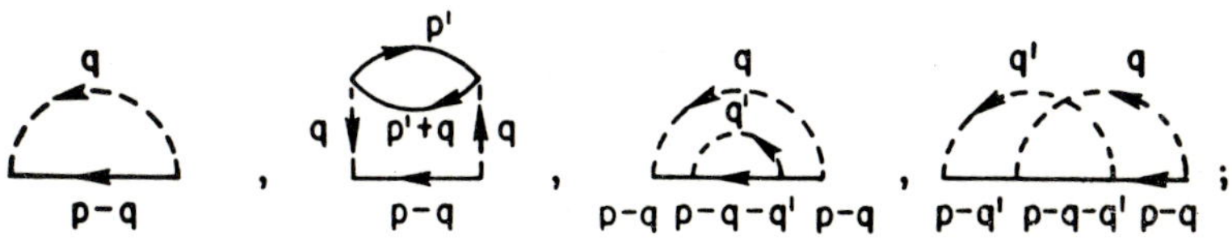

* In this section, we'll assume $w(\mathbf{r}) \equiv 0$.

whereas the following subdiagrams are not:

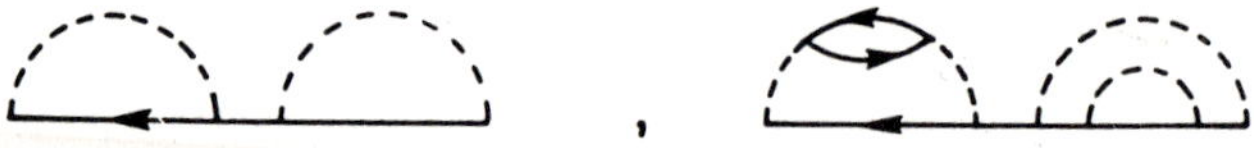

Subdiagrams having the structure of the shaded bump we shall call *proper self-energy* subdiagrams—"proper" because the subdiagram can't be divided cutting only one solid line, "self-energy" for reasons that will soon be apparent. Now every Green function diagram beyond the zeroth order is the product of three contributions: $iG_0(p)$ from the first part, the contribution of a proper self-energy subdiagram for the second part, and the contribution of the third part which, you will see, looks like a Green function diagram (having one line in and one line out). We get *all* Green function diagrams by summing over all proper self-energy subdiagrams in the second factor and over all Green function diagrams in the third factor and adding iG_0 from the zeroth order diagram:

$$iG(p) = iG_0 + [iG_0(p)] \times [\text{Sum over all proper self-energy subdiagrams with total four-momentum } p] \times [\text{sum over all Green function diagrams}],$$

or collecting factors of i,

$$G(p) = G_0(p) + G_0(p)\Sigma(p)G(p). \tag{3.53}$$

Here, $-i\Sigma(p)$ is the sum over all proper self-energy subdiagrams. We shall call $\Sigma(p)$ the *proper self-energy part.*

Equation (3.53), which is Dyson's equation, is represented diagrammatically then as follows:

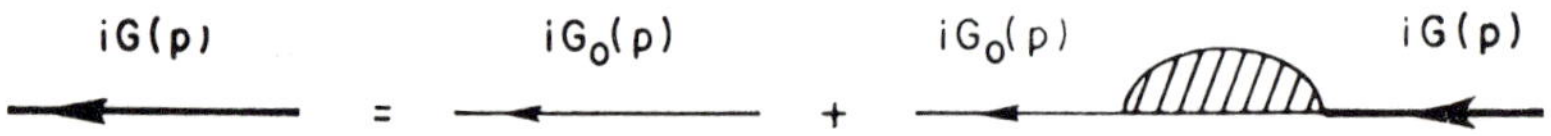

Since all the quantities appearing are just c-number functions of p (an advantage of p-space over x-space, where we get an integral equation), we can solve for $G(p)$ in terms of $\Sigma(p)$:

$$G(p) = \frac{1}{G_0^{-1}(p) - \Sigma(p)}. \tag{3.53'}$$

The problem then is to find $\Sigma(p)$ which, as claimed, is a sum over many fewer diagrams (only proper self-energy subdiagrams) than is $G(p)$.

To see why $\Sigma(p)$ is called a "self-energy" part, recall the expression for $G_0(p)$:

$$G_0(p) = \frac{1}{E-\epsilon_0(\mathbf{p})+i\delta_p}. \tag{3.54}$$

Substituting in (3.53′), we have for $G(p)$:

$$G(p) = \frac{1}{E-\{\epsilon_0(\mathbf{p})+\Sigma(p)\}+i\delta_p}. \tag{3.53″}$$

Comparing these two expressions, we see that $\Sigma(p)$ plays the role of the correction to the zeroth-order one-particle energy to give the "true" one-particle energy, and that is why it is called the self-energy. Actually, this is an oversimplification, because $\Sigma(p)$ depends on E as well as $\mathbf{p}$, so that $G(p)$ does not have the simple structure of $G_0(p)$. Nevertheless, as we have seen, the poles of the analytic continuation of G define the energies and damping of the single particle elementary excitations, so that $\Sigma(p)$ evaluated at the pole, will give the real and imaginary parts of the self-energy.

In general $\Sigma(p)$ will be complex, even for real E, so the factor $i\delta_p$ will usually be omitted. Writing $\Sigma(p) = \Delta(p)+i\Gamma(p)$ for E just above the real axis, it is then clear (from the general property of G that $G^*(p, E-i\eta) = G(p, E+i\eta)$) that $\Sigma = \Delta-i\Gamma$ for E just below the real axis. $\Sigma(p)$ for real E will be the limit as approached from above the real axis when $E > \mu$, the limit as approached from below the real axis when $E < \mu$.

It may happen that we wish to consider diagrams other than one-particle Green function diagrams. In this case, a great reduction in the number of diagrams can be achieved in principle if we consider only diagrams in which no particle lines carry self-energy parts, i.e. so-called self-energy *irreducible* diagrams. The contribution of the full set of diagrams is recovered if to each particle line in each irreducible diagram we assign $iG(p)$ rather than $iG_0(p)$. This is usually too ambitious because we do not know $iG(p)$. A step in this direction is still possible if we consider only diagrams in which any of a definite small class of proper self-energy subdiagrams are absent from every particle line (i.e. partially reduced diagrams) and then use for every particle line $iG_a \equiv i(G_0{}^{-1} - \Sigma_a)^{-1}$, where $-i\Sigma_a$ is the sum over the small class of proper self-energy subdiagrams. Thus, Σ_a might include only  and then we would consider only diagrams in which this subdiagram is always absent. This is exactly equivalent to replacing each particle line  by the infinite sum

a very useful device.

It is instructive to look explicitly at the self-energy part Σ_a which was found in Section F8 of this chapter. If we set $\eta = 0$, we have

$$\Sigma_a(p) = \tfrac{1}{2}\int \frac{d^3q}{(2\pi)^3}\, v(\mathbf{q}) - \int \frac{d^3q}{(2\pi)^3} f_{p-q}\, v(\mathbf{q}).$$

The first term is just $\frac{1}{2}v(\mathbf{r} = 0)$ and cancels the infinite self-energy already lumped in with the single particle energy at the end of Chapter I. The second term in $\Sigma_a(p)$ is just the exchange interaction energy of a particle of momentum $\mathbf{p}$ with the Fermi sea of other particles.

2. *Exact Two-Body Interaction*

We can pull exactly the same trick on the interaction lines that we have just pulled on the particle lines. Suppose we have a diagram with an interaction line in it somewhere. Then in summing over all diagrams, we will encounter diagrams exactly the same as this one with the sole exception that the interaction line is replaced by something more complicated, beginning and ending however with an interaction line. We can sum a large subset of all diagrams if we sum over all subdiagrams that can replace the single interaction line, while holding the rest of the diagram constant. This means that we replace the simple interaction line  by a line  which is a sum constructed in complete analogy to the construction of G from G_0 and Σ in the preceding section:

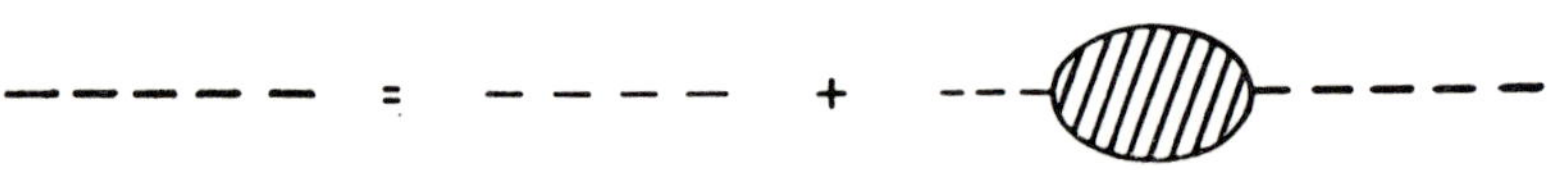

Here, $i\Lambda(q)$ is a sum over all ***proper polarization subdiagrams*** [if $v(q)$ is Coulombic, one might call these proper longitudinal photon self-energy subdiagrams, to stress the analogy with quantum electrodynamics], the first few diagrams being:

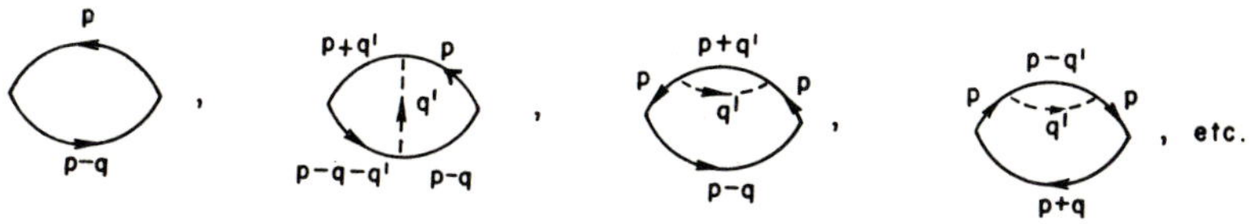

The diagrams are proper in the sense that  cannot be split into two parts by cutting only one interaction line. A diagram which is excluded from $i\Lambda(q)$ because it is improper is

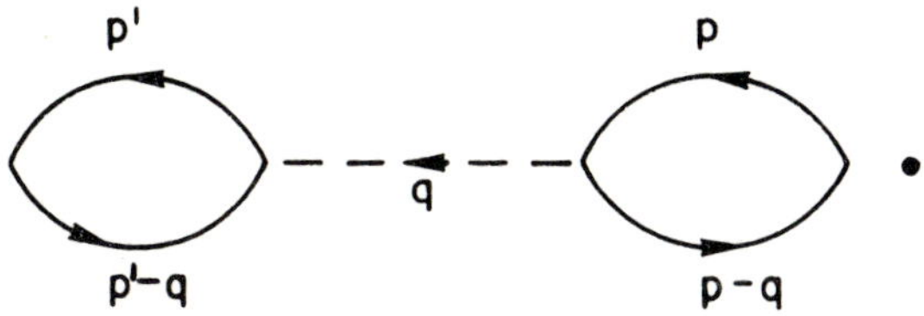

We are thus led to the equation

$$\mathscr{V}(q) = v(q)+v(q)\Lambda(q)\mathscr{V}(q) \tag{3.55}$$

having the solution

$$\mathscr{V}(q) = \frac{v(q)}{1-v(q)\Lambda(q)}. \tag{3.55'}$$

It will become clear that to include diagrams contributing to $\Lambda(q)$ is to include the polarization of the Fermi sea in the interaction $\mathscr{V}(q)$ between particles; hence the name polarization part. Observe that $v(q)$ and $\mathscr{V}(q)$ play the role of Green functions or propagators in that they are attached to the zeroth order and exact interaction lines.

The introduction of $\mathscr{V}(q)$ also allows a reduction in the number of diagrams to be drawn. We draw only diagrams having no proper polarization parts in them (so-called polarization irreducible diagrams), and to each interaction line associate the factor $-i\mathscr{V}(q)$— which, of course, we don't know but try to compute by trying to find $\Lambda(q)$. Because $\mathscr{V}(q)$ is the interaction including all polarization effects, the relation (3.55′) suggests that we define the generalized dielectric constant $\kappa(q)$:

$$\mathscr{V}(q) = \frac{v(q)}{\kappa(q)} \Rightarrow \kappa(q) = 1 - v(q)\Lambda(q) \tag{3.56}$$

This dielectric constant, and another one closely associated with it, play an important role in the many-body problem. In Section I we'll go into a more detailed discussion of the whole subject of the dielectric constant. Notice that, though $v(q)$ does not depend on ν, reflecting the fact that $v(x)$ is instantaneous, $\mathscr{V}(q)$ does depend on ν, reflecting the fact that the polarization charges which contribute to the screening have inertia. The interaction $\mathscr{V}(x)$ is not instantaneous.

3. *Ladder Sums*

A third subset of diagrams that can be summed corresponds to repeated scatterings of two particles on one another or of one particle on an external potential. The effect is to replace the first Born approximation for the scattering amplitude (the Fourier transform of the potential) by the exact scattering amplitude, taking into account the exclusion-principle effects of the Fermi sea. This is particularly important when the first Born approximation is divergent as is the case for a hard core potential (e.g., nuclear matter or liquid helium). It is also useful if, in addition to scattering states, there exist bound states of two particles or of one particle in the external potential (e.g., excitons, bound impurity electrons). For example, the following series of diagrams would give contributions to the electron self-energy $\Sigma(p)$ due to the creation of virtual excitons (bound electron-hole pairs) in intermediate states:

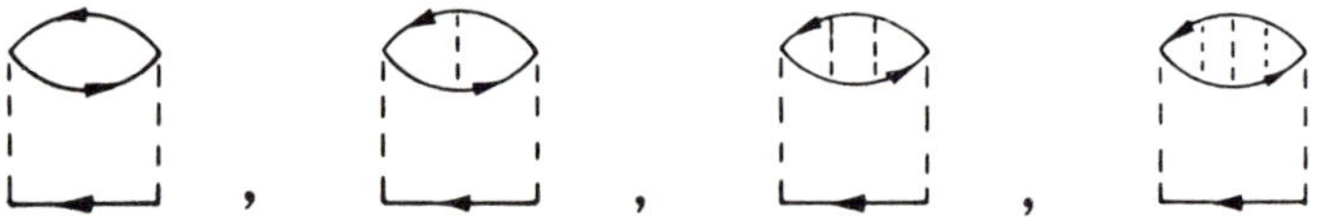, etc.

The argument for the partial summation proceeds as in the preceding two sections. For simplicity let us consider only the repeated scattering of a particle by an external source, $w(\mathbf{r})$. For any diagram in which there is an interaction with an external potential, there are other diagrams identical in all respects with this one except that the single interaction is replaced by a repeated interaction. We are led to consider the sum of all subdiagrams of the following form:

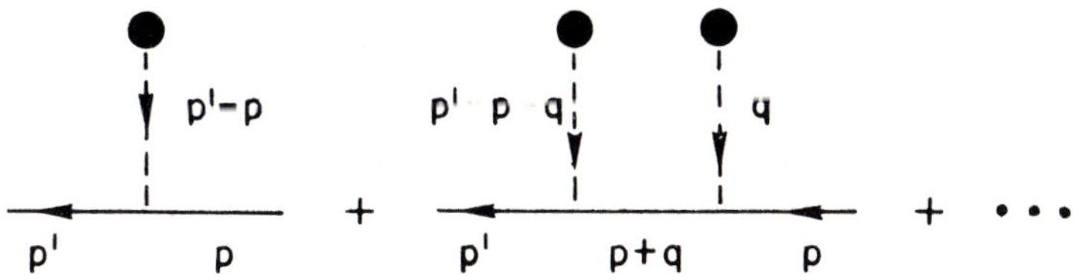

To sum these, we get an integral equation from the following schematic picture:

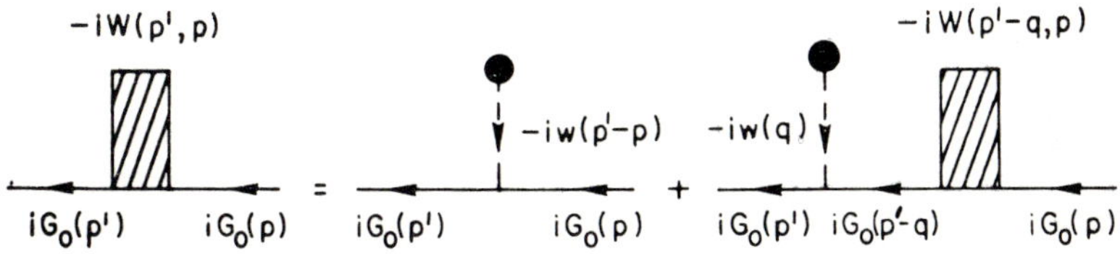

Collecting factors of i and omitting $G_0(p)$ and $G_0(p')$ the integral equation is,

$$W(p', p) = w(p'-p)+\int \frac{d^4q}{(2\pi)^4}\, w(q)G_0(p'-q)W(p'-q, p). \tag{3.57}$$

Now if $w(x)$ is time independent,

$$w(q) = 2\pi w(\mathbf{q})\delta(\nu).$$

Then we can do the ν integration obtaining

$$W(p', p) = w(\mathbf{p}'-\mathbf{p}) \,.\, 2\pi\delta(E'-E) + \int \frac{d^3q}{(2\pi)^3}\, w(\mathbf{q})G_0(\mathbf{p}'-\mathbf{q}, E')W(p'-q, p).$$

By inspection, $W(p', p)$ is proportional to $\delta(E'-E)$,

$$W(p', p) \equiv W(\mathbf{p}', \mathbf{p};\, E) \cdot 2\pi\delta(E'-E),$$

and $W(\mathbf{p}', \mathbf{p}, E)$ satisfies

$$W(\mathbf{p}', \mathbf{p};\, E) = w(\mathbf{p}'-\mathbf{p}) + \int \frac{d^3q}{(2\pi)^3}\, w(\mathbf{q})G_0(\mathbf{p}'-\mathbf{q}, E)W(\mathbf{p}'-\mathbf{q}, \mathbf{p};E). \tag{3.58}$$

This is just the integral equation for the exact scattering amplitude for particles of energy E and incident momentum $\mathbf{p}$ on the unscreened potential $w(\mathbf{p})$. The use of G_0 on the right insures inclusion of the effects of the Fermi sea via the exclusion principle.

If we associate the factor $W(p', p)$ rather than $w(p'-p)$ with each interaction, we should then sum only those diagrams having no multiple scatterings like 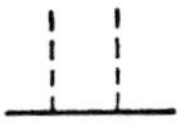 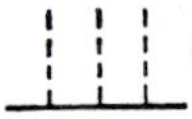 etc. A much larger class or diagrams is included if instead of $G_0(p)$ and $v(q)$ one uses some bettet approximations to $G(p)$ and $\mathscr{V}(q)$, thus including the effects on self-energy and screening due to the Fermi sea. Note however that this generalization is still far from a description of the exact effects of the potential $w(q)$ on $G(p', p)$. To describe these one would have to include diagrams like

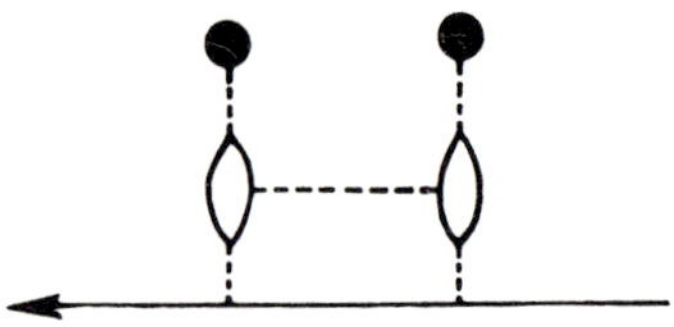

in which successive scatterings are not independent, and diagrams like

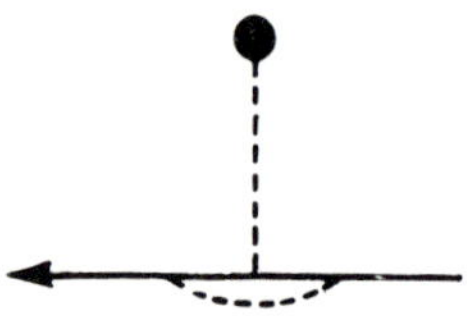 , 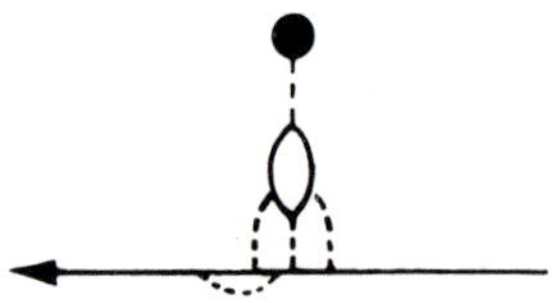

in which the simple vertex  is replaced by more complex "vertex subdiagrams".

A more detailed discussion of ladder sums for the scattering of two particles, as used in the theory of nuclear matter, is given by Galitskii.[6]

In conclusion then, we have seen at least three kinds of partial summations that are possible in diagrams. To the extent that we can construct G instead of G_0, $\mathscr{V}$ instead of v, and W instead of w, we are encouraged to draw only those diagrams in which particle lines have no self-energy parts, interaction lines have no polarization parts, and scatterings from $w(q)$ are never multiple, associating with the lines iG and $-i\mathscr{V}$ respectively and with the scatterings $-iW(p, p')$ or its generalization for two-body scattering. In practice we can never determine G, or $\mathscr{V}$, or even W exactly. Also, some partial sums may not be so important as others. The polarization sum is more important in the electron gas in metals; the ladder sum is more important for liquid helium or nuclear matter. Which of these sums, or other possible partial sums, we perform depends on the particular problem. It is important now only to realize that such sums are possible, important, and often necessary (e.g., to get any damping of single-particle excitations, one must sum an infinite number of self-energy subdiagrams; to get rid of the infinities arising from the long-range Colomb interaction, one must sum infinitely many polarization subdiagrams; to avoid the divergent Fourier transform of a hardcore potential, one should sum all ladder diagrams. In all these cases, the formal sums of apparently divergent diagrams yield finite and meaningful results).

H. EXPRESSIONS FOR THE GROUND STATE ENERGY OF A FREE ELECTRON GAS

1. *In Terms of the Dielectric Constant*

Recall expression (3.9) for the total energy in terms of the potential energy for all values of the charge, $e' < e$:

$$\mathscr{E} = \mathscr{E}_0 + \int_0^{e^2} \frac{de'^2}{e'^2}\Big[\langle\Psi_0(e'^2)|H_1(e'^2)|\Psi_0(e'^2)\rangle - \tfrac{1}{2}Nv_{e'}(\mathbf{r} = 0)\Big]. \tag{3.59}$$

Thus we must find $\langle\Psi_0|H_1|\Psi_0\rangle$ for $0 < e'^2 < e^2$. Now

$$\begin{aligned}\langle\Psi_0|H_1|\Psi_0\rangle &= \tfrac{1}{2}\int d^4x_1\, d^4x_2\, \delta(t_1)v(x_1-x_2)\langle\Psi_0|\bar{\psi}(x_1)\psi(x_1)\bar{\psi}(x_2)\psi(x_2)|\Psi_0\rangle \\ &= -\tfrac{1}{2}\int d^4x_1\, d^4x_2\, v(\mathbf{r}_1-\mathbf{r}_2)\delta(t_1-t_2{}^{++})\delta(t_1) \\ &\qquad G(\mathbf{r}_1t_1, \mathbf{r}_2t_2; \mathbf{r}_1t_1{}^+, \mathbf{r}_2t_2{}^+).\end{aligned} \tag{3.60}$$

Also $-\frac{1}{2}G(x_1x_2;\, x_1'x_2')$ can be represented diagrammatically:

$$-\tfrac{1}{2}G(x_1x_2;x_1'x_2') = -\tfrac{1}{2}(-i)^2 \times \tag{3.61}$$

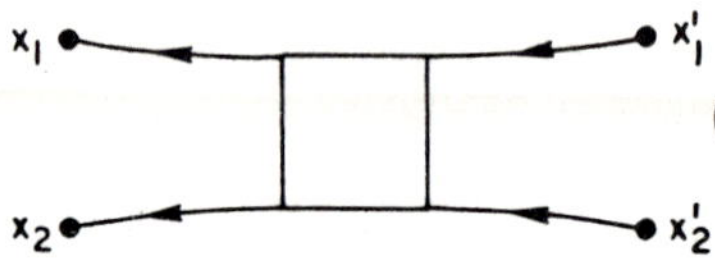

The box means the sum over all subdiagrams like

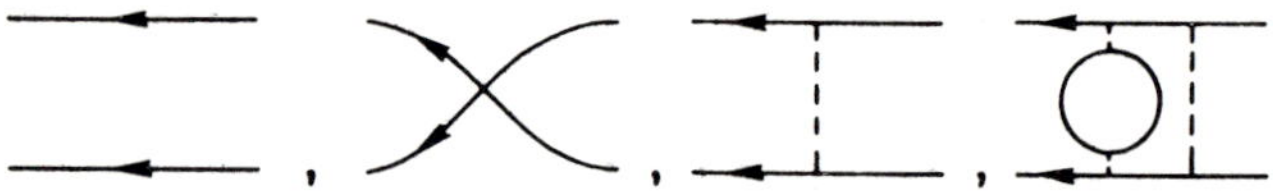

etc.

Equation (3.60) in diagrams is

$$\langle \Psi_0 | H_1 | \Psi_0 \rangle = \tfrac{1}{2}\left(\frac{1}{-i}\right) \tag{3.62}$$

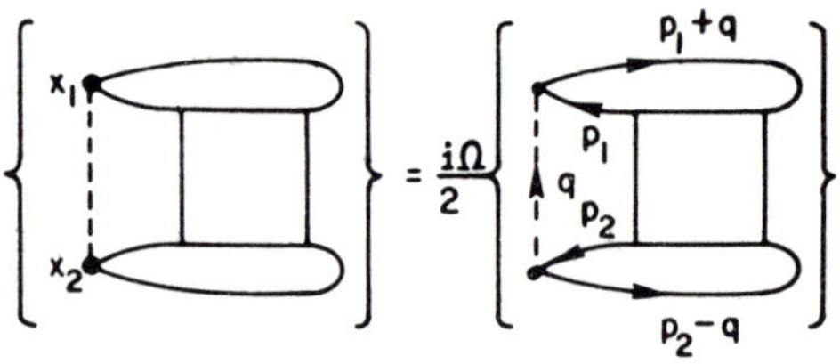

where t_1 is fixed at $t_1 = 0$.

First note that the simplest structure for (3.62) gives zero:

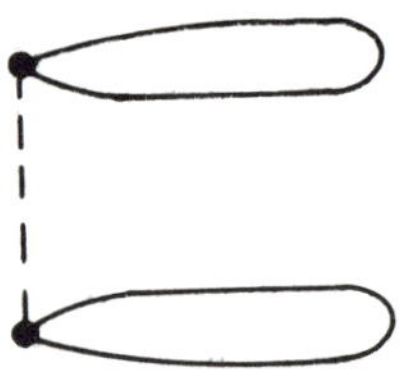

All other diagrams in (3.62) can be put into one of two classes:

Class I: Those in which the box part of the diagram cannot be divided into two parts by breaking only one dotted line, and

Class II: Those in which the box can be so divided. In this case draw the diagram thusly:

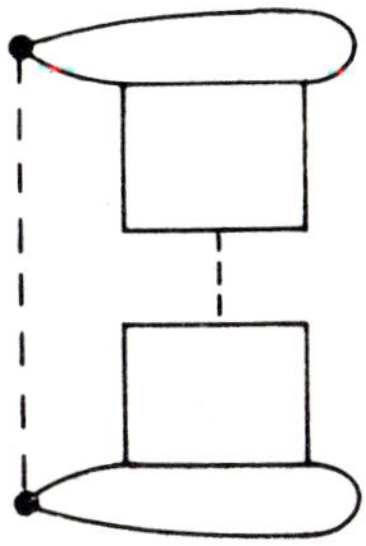

where the lower box cannot be further divided into two, breaking only one dotted line.

The sum of all diagrams of Class I can be represented by the diagram

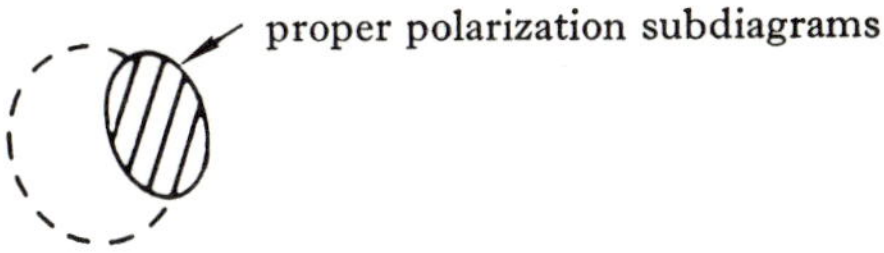

and the sum of all diagrams of Class II can be represented by

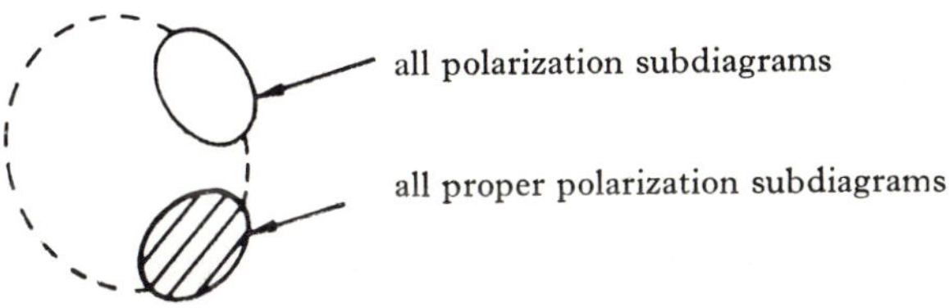

Adding these two together, we have for the diagrams of (3.62),

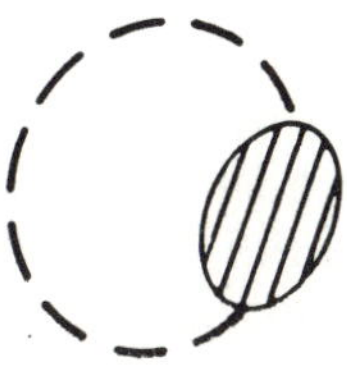

$$= \int \frac{d^4q}{(2\pi)^4}\{-i\mathscr{V}(q)\}\{i\Lambda(q)\},$$

so that

$$\langle\Psi_0|H_1|\Psi_0\rangle = \frac{i\Omega}{2}\int \frac{d^4q}{(2\pi)^4}\mathscr{V}(q)\Lambda(q). \tag{3.63}$$

The integration on ν can be performed by closing the contour; although the integral on the large semi-circle will not always vanish, it can be evaluated for low order contributions to the integrand. One must be careful not to interchange the order of the ν-integration with any of the energy integrations involved in evaluating $\mathscr{V}(q)$ or $\Lambda(q)$ however. As you will readily see in lowest order, where

$$\mathscr{V}(q) \simeq v(q) \quad \text{and} \quad \Lambda(q) \simeq i\int \frac{d^4p}{(2\pi)^4} G_0(p-q)G_0(p),$$

there is not always the required uniform convergence.

With this warning, let us write (3.63) in terms of the generalized dielectric constant alone. From

$$\mathscr{V}(q) = \frac{v(q)}{\kappa(q)} \quad \text{and} \quad \kappa(q) = 1-v(q)\Lambda(q)$$

we obtain

$$\mathscr{V}(q)\Lambda(q) = \frac{1}{\kappa(q)}-1, \tag{3.64}$$

so that

$$\langle\Psi_0|H_1|\Psi_0\rangle = \frac{i\Omega}{2}\int \frac{d^4q}{(2\pi)^4}\left(\frac{1}{\kappa(q)}-1\right).$$

But the left side must be real, since H_1 is Hermitian. Therefore,

$$\langle\Psi_0|H_1|\Psi_0\rangle = -\frac{\Omega}{2}\int \frac{d^4q}{(2\pi)^4} Im\frac{1}{\kappa(q)} \tag{3.65}$$

which, with (3.59), gives the ground state energy in terms of $\kappa(q)$ for all $e'^2 < e^2$. This result has been derived by the use of diagrams. The result is actually more general as shown by Nozières and Pines.[7] In the following section we shall discuss the dielectric constant in more detail and show this relation directly, without perturbation theory.

The expression for the energy in terms of the dielectric constant proves very useful for the high density electron gas, where it is possible to obtain good approximations to the dielectric constant relatively easily, as discussed in section J of this chapter.

2. *In Terms of the Proper Self-Energy Part*

Here's another way to classify the diagrams having the structure of (3.62), which leads to a different expression for the ground state energy, this time in terms of the proper self-energy part, $\Sigma(p)$. Such an expression may be a better starting point for an approximation procedure than (3.65).

All diagrams of the structure

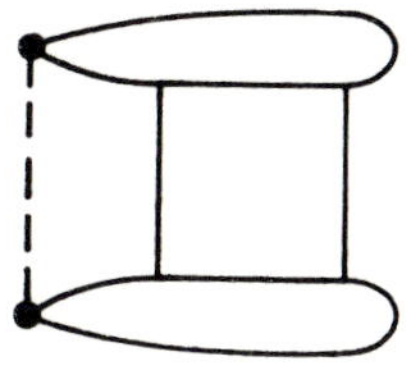

can be put into one of two classes:

Class I: Those in which the lower right corner of the box cannot be separated from the rest of the box breaking only one solid line, and

Class II: Those in which the box can be so divided. In this case, separate off the largest corner possible breaking only one solid line.

A little study will show that the sum of all diagrams of Class I can be represented by the diagram

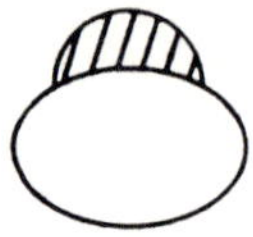

while the sum of all diagrams of Class II can be represented by

all self-energy subdiagrams 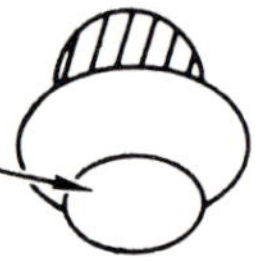

Adding these two together we have

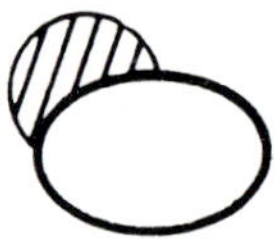

leading to the equation

$$\langle\Psi_0|H_1|\Psi_0\rangle = \frac{i\Omega}{2}(-1)\int \frac{d^4p}{(2\pi)^4}\{-i\Sigma(p)\}\{iG(p)\},$$

the (-1) coming from the one closed loop additional to those in .

This expression as it stands is not quite right. The contribution to $\Sigma(p)$ from the subdiagram 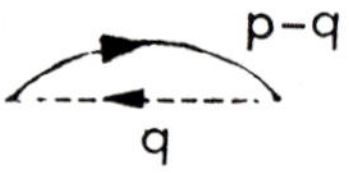

is

$$\tfrac{1}{2}i\int \frac{d^4q}{(2\pi)^4}\, v(q)G_0(p-q)(e^{i\nu\eta}+e^{-i\nu\eta})$$

as we discussed in Section F8d of this chapter, whereas what is wanted here is

$$i\int \frac{d^4q}{(2\pi)^4}\, v(q)G_0(p-q)$$

corresponding to the fact that in diagram of (3.62) $t_1 = t_2+\eta$. $\Sigma(p)$ with this modification we shall call $\Sigma_+(p)$. The modification only effects the lowest order contribution to $\Sigma(p)$ and does not change its value when considered as a contribution to the single-particle self-energy. It does insure cancellation of the infinite self-energy term $-\frac{1}{2}Nv(\mathbf{r}=0)$, however.

Because the left side is real, we may finally write

$$\langle\Psi_0|H_1|\Psi_0\rangle = -\frac{i}{2}\int\frac{d^4p}{(2\pi)^4}\,\Sigma_+(p)G(p) = \tfrac{1}{2}Im\int\frac{d^4p}{(2\pi)^4}\Sigma_+(p)G(p) \tag{3.66}$$

which, if known for all $e'^2 < e^2$, can be used to determine $\mathscr{E}_0$.

If we recall that

$$n(\mathbf{p}) = \frac{-i\Omega}{(2\pi)^3}\int_{-\infty}^{\infty}\frac{dE}{2\pi}\,e^{iE\eta}\,G(p),$$

then it is obvious that

$$\langle\Psi_0|H_0|\Psi_0\rangle = \int\frac{d^3p}{(2\pi)^3}n(\mathbf{p})\epsilon(\mathbf{p}) = -i\Omega\int\frac{d^4p}{(2\pi)^4}e^{iE\eta}\epsilon(\mathbf{p})G(p). \tag{3.67}$$

Combining these, we obtain

$$\langle\Psi_0|H|\Psi_0\rangle = -i\Omega\int\frac{d^4p}{(2\pi)^4}e^{iE\eta}\left[\epsilon(\mathbf{p}) + \frac{\Sigma_+(p)}{2}\right]G(p). \tag{3.68}$$

Although $\Sigma_+(p)$ is the self-energy of a particle of momentum $\mathbf{p}$ [provided E is chosen properly in $\Sigma_+(p)$] we see that the total energy includes only half the self-energy for each particle. This is because the self-energy of one particle involves the interaction energy with all the other particles. To add all the self-energies together would be to include all interaction energies twice. Of course, it is really $\Sigma_+(p)$ for arbitrary values of $(\mathbf{p}, E)$ which enters in (3.68) and this is a bit more generalized notion than this simple intuitive interpretation.

It is also possible to derive (3.66) without recourse to diagrams or perturbation theory, but rather from the Heisenberg equations of motion for the various operators $C_p(t)$. The Heisenberg equation for $C_p(t)$ is

$$\begin{aligned}\frac{\partial C_p(t)}{\partial t} &= i[H, C_p(t)]\\ &= i\Big[\sum_{p'}\epsilon_0(\mathbf{p}')C_{p'}^{\dagger}C_{p'} + \tfrac{1}{2}\sum_{p'p''q}v(\mathbf{q})C_{p'}^{\dagger}C_{p''}^{\dagger}C_{p''+q}C_{p'-q},\,C_p\Big]\\ &= -i\Big(-\epsilon_0(\mathbf{p})C_p - \sum_{p',q}v(\mathbf{q})C_{p'}^{\dagger}C_{p'-q}C_{p+q}\Big),\end{aligned} \tag{3.69}$$

where we have assumed

$$v(\mathbf{q}) = v(-\mathbf{q})$$

to get the second line. All operators are assumed to be at the time t. If we multiply by $(i/2)C_p{}^\dagger(t)$ from the left and sum on $\mathbf{p}$, rearranging terms, we obtain

$$H_1' = \tfrac{1}{2}\sum_p \left(iC_p{}^\dagger \frac{\partial C_p}{\partial t} - \epsilon_0(\mathbf{p})C_p{}^\dagger C_p\right). \tag{3.70}$$

Here, $H_1' = H_1 - \frac{1}{2}Nv(\mathbf{r}=0)$ is the original potential energy. Now recall the definition

$$G(\mathbf{p}, t) = i\langle \Psi_0 | T(C_p{}^\dagger(0)C_p(t))|\Psi_0\rangle.$$

Differentiating at $t = -\eta$, η a small positive infinitesimal we obtain

$$\left[\frac{\partial G(\mathbf{p}, t)}{\partial t}\right]_{t=-\eta} = i\left\langle \Psi_0 \middle| C_p{}^\dagger(0)\left[\frac{\partial C_p(t)}{\partial t}\right]_{t=-\eta}\middle|\Psi_0\right\rangle. \tag{3.71}$$

Combining (3.70) and (3.71), we have $\langle\Psi_0|H_1'|\Psi_0\rangle$ expressed in terms of $G(\mathbf{p}, t)$ alone:

$$\begin{aligned}\langle\Psi_0|H_1'|\Psi_0\rangle &= \tfrac{1}{2}\sum_p \left[\left(\frac{\partial}{\partial t} + i\epsilon_0(\mathbf{p})\right)G(\mathbf{p}, t)\right]_{t=-\eta} \\ &= \frac{\Omega}{2}\int \frac{d^3p}{(2\pi)^3}\int dt\, \delta(t+\eta)\left(\frac{\partial}{\partial t} + i\epsilon_0(\mathbf{p})\right)G(\mathbf{p}, t).\end{aligned}$$

Introducing the Fourier integral representation of the delta function and interchanging orders of integration, we get

$$\langle\Psi_0|H_1'|\Psi_0\rangle = \frac{\Omega}{2}\int \frac{d^4p}{(2\pi)^4}\int dt\, e^{iE(t+\eta)}\left(\frac{\partial}{\partial t} + i\epsilon_0(\mathbf{p})\right)G(\mathbf{p}, t).$$

Integrating by parts,

$$\langle\Psi_0|H_1'|\Psi_0\rangle = -\frac{i\Omega}{2}\int \frac{d^4p}{(2\pi)^4}(E - \epsilon_0(\mathbf{p}))\, e^{iE\eta}\int dt\, e^{iEt}G(\mathbf{p}, t),$$

or

$$= -\frac{i\Omega}{2}\int \frac{d^4p}{(2\pi)^4}\, G_0{}^{-1}(p)G(p)\, e^{iE\eta}. \tag{3.72}$$

But we can now *define* $\Sigma(p)$ by the Dyson equation

$$G = G_0 + G_0 \Sigma G. \tag{3.53}$$

(If we do not use diagrams, then we *need* an alternative definition of Σ. To be consistent with the diagram approach, we define $\Sigma(p)$ by (3.53).] From the Dyson equation we obtain

$$G_0^{-1}(p)\, G(p) = 1 + \Sigma(p) G(p)$$

so that

$$\langle \Psi_0 | H_1' | \Psi_0 \rangle = -\frac{i\Omega}{2} \int \frac{d^4p}{(2\pi)^4} e^{iE\eta} \left(1 + \Sigma(p) G(p)\right).$$

But

$$\int \frac{d^4p}{(2\pi)^4} e^{iE\eta} \propto \delta(\eta) = 0$$

because $\eta \neq 0$. Therefore we finally get (3.66), aside from differences involving the self-energy $\frac{1}{2} N v(\mathbf{r} = 0)$ (which are connected with the difference between H_1 and H_1' and with the definition of Σ, which we know must cancel).

I. THE DIELECTRIC CONSTANT AND THE POLARIZATION PROPAGATOR

In Section G we were led to a generalized dielectric constant $\kappa(q)$ whose reciprocal was essentially the polarization propagator, i.e., it played the same role in relation to interaction lines as the Green function plays for particle lines. We also obtained an expression for the energy in terms of this dielectric constant. It turns out that the dielectric constant is a fundamental quantity in the description of a many-body system, that it can be analyzed in considerable detail without perturbation theory, and several exact results can be obtained. The dielectric constant as discussed by Nozières and Pines[7] is slightly different from the one we have introduced, so I should like to discuss both of them and their relation to one another.

1. *Retarded Commutator Polarization Propagator*

Suppose we introduce an external space-time-dependent charge distribution into our many-body system with number density

$\rho_{\text{ex}}(\mathbf{r}t)$.* Suppose it has the Fourier representation

$$\rho_{\text{ex}}(\mathbf{r}t) = \int \frac{d^4q}{(2\pi)^4} \rho_{\text{ex}}(\mathbf{q}\nu)\, e^{i(\mathbf{q}\cdot\mathbf{r}-\nu t)}. \tag{3.73}$$

Now let us calculate the expectation value of the number density of induced charges:

$$\begin{aligned} \rho_{\text{in}}(\mathbf{r}t) &= \langle \Psi(t)|\bar{\psi}(\mathbf{r}t)\psi(\mathbf{r}t)|\Psi(t)\rangle - \langle \Psi_0|\bar{\psi}(\mathbf{r}t)\psi(\mathbf{r}t)|\Psi_0\rangle \\ &= \langle \Psi(t)|\rho(\mathbf{r}t)|\Psi(t)\rangle - \langle \Psi_0|\rho(\mathbf{r}t)|\Psi_0\rangle, \end{aligned} \tag{3.74}$$

where $|\Psi_0\rangle$ is the exact ground state in the absence of any external charge, $|\Psi(t)\rangle$ is the state developing from $|\Psi_0\rangle$ as we turn on the external charge, and $\rho(\mathbf{r}t)$ is the number density operator. Here we consider the full many-body Hamiltonian. Then the operators $\psi(x)$ and $\bar{\psi}(x)$ develop in time according to H, and the state vector $\Psi(t)$ develops according to $H'(t)$ [see (3.28′), (3.29), and (3.33)].

Assuming the external charge to be small, we seek the linear response of the system to this charge. The most general linear relation between $\rho_{\text{ex}}(x)$ and $\rho_{\text{in}}(x)$ for a spatially homogeneous time-independent unperturbed system is

$$\rho_{\text{in}}(x) = \int d^4x' L_r(x-x')\rho_{\text{ex}}(x').$$

We might seek the kernel $L_r(x)$; however it is customary to consider instead the closely related polarization propagator $P_r(x)$, which is the kernel expressing the linear relation between $\rho_{\text{in}}(x)$ and the potential field produced by $\rho_{\text{ex}}(x)$. Thus $P_r(x)$ is defined by

$$\rho_{\text{in}}(x) = \int d^4x' P_r(x-x') \int d^4x'' v(x'-x'')\rho_{\text{ex}}(x''). \tag{3.75}$$

If we introduce the three-momentum Fourier transforms of $\rho_{\text{ex}}(x)$, $\rho_{\text{in}}(x)$, and $P_r(x)$, we have

$$\rho_{\text{in}}(\mathbf{q}t) = \int dt' P_r(\mathbf{q}, t-t') \int dt'' v(\mathbf{q}, t'-t'')\rho_{\text{ex}}(\mathbf{q}t'') \tag{3.75′}$$

* We have in mind a system of point charges for which $v(q)$ is the Coulomb interaction and so lapse into the appropriate terminology, although the development of this section, not making use of this fact, has broader application.

and if we introduce the four-dimensional Fourier transforms we have

$$\rho_{\text{in}}(q) = P_r(q)v(q)\rho_{\text{ex}}(q) \tag{3.75''*}$$

But a generalized dielectric constant would be defined by

$$\rho_{\text{total}}(q) \equiv \rho_{\text{ex}}(q)+\rho_{\text{in}}(q) = \frac{\rho_{\text{ex}}(q)}{\kappa_r(q)}, \text{ or } \rho_{\text{in}}(q) = \rho_{\text{ex}}(q)\left(\frac{1}{\kappa_r(q)}-1\right)$$

so that the polarization propagator and dielectric constant are related by

$$v(q)P_r(q) = \frac{1}{\kappa_r(q)}-1. \tag{3.76}$$

Now let us calculate $P_r(q)$ by perturbation theory on H', but treating all interactions *among* the particles of our system exactly. The interaction operator at the time t' can be written

$$H'(t') = \int d^3r'\, d^4x''\, \rho(\mathbf{r}'t')v(x'-x'')\rho_{\text{ex}}(x''),$$

where it is understood that $v(x'-x'')$ may be proportional to $\delta(t-t'')$. To first order in H', the state vector $\Psi(t)$ is

$$\Psi(t) = \left[1-i\int_{-\infty}^{t} H'(t')dt'\right]\Psi_0.$$

Substituting in (3.74) and neglecting second-order terms, we have

$$\rho_{\text{in}}(\mathbf{r}t) = i\int_{-\infty}^{t} dt'\int d^3r'\,\langle\Psi_0|[\rho(\mathbf{r}'t'),\, \rho(\mathbf{r}t)]|\Psi_0\rangle\int d^4x''\, v(x'-x'')\rho_{\text{ex}}(x'').$$

Comparing this with (3.75′), we see that

$$\begin{aligned} P_r(x-x') &= i\langle\Psi_0|[\rho(\mathbf{r}'t'),\, \rho(\mathbf{r}t)]|\Psi_0\rangle,\ t' < t, \\ &= 0,\ t' > t. \end{aligned} \tag{3.77}$$

* The definition of the polarization propagator used here differs in some ways from that used by DuBois, for whom the polarization propagator is essentially the proper polarization part $\Lambda(q)$. $P_r(q)$ as used here also differs from DuBois' $P_{r_s}(q,\ \omega)$, which is essentially our $\mathscr{V}(q)$.

Taking the three-momentum transform*, and using the time independence of the system, we find

$$P_r(\mathbf{q},\tau) = i\Omega^{-1}\langle\Psi_0|[\rho(-\mathbf{q}, -\tau), \rho(\mathbf{q}0)]|\Psi_0\rangle, \ \tau > 0,$$
$$= 0, \ \tau < 0. \tag{3.77'}$$

In some ways this looks like a Green function, except that what appears on the right is the matrix element of a commutator rather than a T-product, and the propagator is zero for negative times. What we have is a so-called "retarded commutator" rather than a T-product. The two are closely related, the retarded commutator having simpler analytic properties, but the T-product being more natural in perturbation theory.

Consider the analytic properties of $P_r(\mathbf{q},\tau)$, or better, of its transform, $P_r(q)$:

$$P_r(q) = \int_0^\infty dt\, e^{i\nu t}\, P_r(\mathbf{q},t).$$

To insure convergence let ν have a small positive imaginary part, i.e. let $\nu \to \nu+i\eta$. Then, expanding in a complete set of intermediate states and carrying out the integrations, just as in deriving (3.15), we obtain

$$P_r(q) = -\,\Omega^{-1}\sum_n\left[\frac{\langle 0|\rho(-\mathbf{q})|n\rangle\langle n|\rho(\mathbf{q})|0\rangle}{\nu-\mathscr{E}_0+\mathscr{E}_n+i\eta} - \frac{\langle 0|\rho(\mathbf{q})|n\rangle\langle n|\rho(-\mathbf{q})|0\rangle}{\nu-\mathscr{E}_n+\mathscr{E}_0+i\eta}\right]. \tag{3.78}$$

Now $\rho(q) = \rho^\dagger(-q)$ so that

$$\langle 0|\rho(\mathbf{q})|n\rangle\langle n|\rho(-\mathbf{q})|0\rangle = |\langle 0|\rho(\mathbf{q})|n\rangle|^2.$$

Assume

$$|\langle 0|\rho(\mathbf{q})|n\rangle|^2 = |\langle 0|\rho(-\mathbf{q})|n\rangle|^2,$$

*We make use of the conservation of momentum exhibited by this system, which ensures that

$$\langle \Psi_0|\rho(\mathbf{q}', -\tau)\rho(\mathbf{q}0)|\Psi_0\rangle = \langle \Psi_0|\rho(-\mathbf{q}, -\tau)\rho(\mathbf{q}0)|\Psi_0\rangle\, \delta_{\mathbf{q}',-\mathbf{q}}$$
$$= \frac{(2\pi)^3}{\Omega}\langle \Psi_0|\rho(-\mathbf{q}, -\tau)\rho(\mathbf{q}0)|\Psi_0\rangle\, \delta(\mathbf{q}'+\mathbf{q}).$$

i.e., inversion symmetry for the unperturbed system. Then

$$P_r(q) = \int_{-\infty}^{\infty} \frac{\mathscr{F}(\mathbf{q}, \nu')}{\nu' - \nu - i\eta} d\nu' = \int_{C_r} \frac{\mathscr{F}(\mathbf{q}, \nu')}{\nu' - \nu} d\nu', \tag{3.78'}$$

where $\mathscr{F}(\mathbf{q}, \nu)$ is defined by

$$\mathscr{F}(\mathbf{q}, \nu)d\nu = -\Omega^{-1} \sum_{\nu < \mathscr{E}_n - \mathscr{E}_0 < \nu + d\nu} |\langle 0|\rho(\mathbf{q})|n\rangle|^2 \text{ for } \nu > 0,$$

$$= -\mathscr{F}(\mathbf{q}, -\nu)d\nu \text{ for } \nu < 0, \tag{3.79}$$

and

C_r = 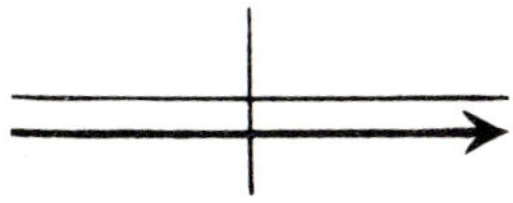

a straight line just below the real axis.

From this Lehmann representation, we see that $P_r(q)$ is analytic in the whole upper half-plane including the real axis, a simpler analytic behavior than we found for the Green functions defined with T-products. The following other properties are now simple to show:

(*a*) $P_r(\mathbf{q}, \nu) = P_r^*(\mathbf{q}, -\nu)$, for ν real. (3.80a)

(*b*) $Im P_r(\mathbf{q}, \nu) = \pi \mathscr{F}(\mathbf{q}, \nu)$, for ν real. (3.80b)

(*c*) $$Re\, P_r(\mathbf{q}, \nu) = \frac{1}{\pi} \mathscr{P} \int_{-\infty}^{\infty} d\nu' \frac{Im\, P_r(\mathbf{q}, \nu')}{\nu' - \nu},$$

and $$P_r(\mathbf{q}, \nu) = \frac{1}{\pi} \int_{-\infty}^{\infty} d\nu' \frac{Im\, P_r(\mathbf{q}, \nu')}{\nu' - \nu - i\eta}. \tag{3.80c}$$

(d) $P_r(\mathbf{q}, \nu)$ goes into its complex conjugate as ν crosses the contour C_r. (3.80d)

2. *Time-Ordered Product Polarization Propagator*

Before deriving various properties of the dielectric constant from the Lehmann representation given in (3.78), let us consider the time-ordered product definition for $P(q)$, the one which agrees with the dielectric constant defined earlier in terms of graphs.

Our original definition of $\kappa(q)$ was through diagrams. In fact, $\kappa(q)$ was defined through $\mathscr{V}(q) = v(q)/\kappa(q)$ and the polarization propagator is then $P_t(q) = [(\kappa_t(q))^{-1} - 1]/v(q)$, where the subscript t is now added to indicate a time-ordered product definition. In diagrams this is

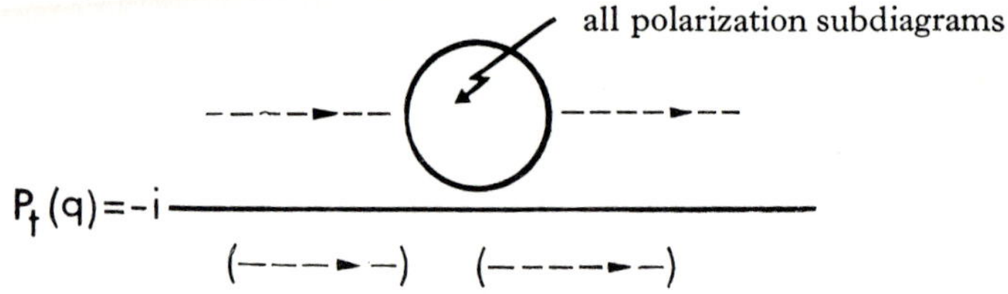

$$= -i \times \text{(all polarization subdiagrams)}.$$

We want to find an operator that would have some matrix element leading to the set of diagrams in the numerator. Consider the ground state expectation value of the interaction twice with an external charge distribution $\rho_{ex}(rt)$:

$$J = \left\langle \Psi_0 \right| \int dt\, dt'\, T \left(\int d^3r_1\, d^3r_2\, v(\mathbf{r}_1 - \mathbf{r}_2)\rho_{ex}(\mathbf{r}_1 t)\rho(\mathbf{r}_2 t) \right.$$
$$\left. \times \int d^3r'_1\, d^3r_2'\, v(\mathbf{r}_1' - \mathbf{r}_2')\rho_{ex}(\mathbf{r}_1' t')\rho(\mathbf{r}_2' t') \right) \left| \Psi_0 \right\rangle.$$

J is given by minus the sum over all diagrams having the following general structure, integrated over $(\mathbf{r}_1 t)$ and $\mathbf{r}_1' t')$.

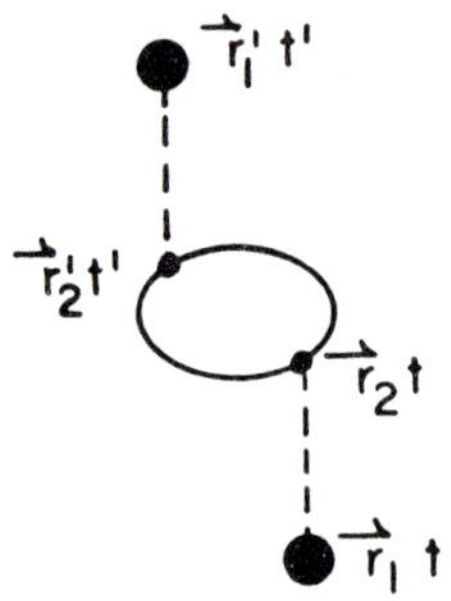

In momentum space

$$J = -\int \frac{d^4q}{(2\pi)^4} \rho_{ex}(q) \left[\text{diagram} \right] \rho_{ex}(-q),$$

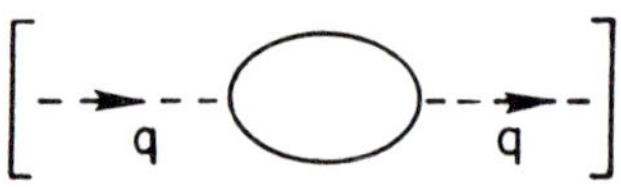

which allows us to identify

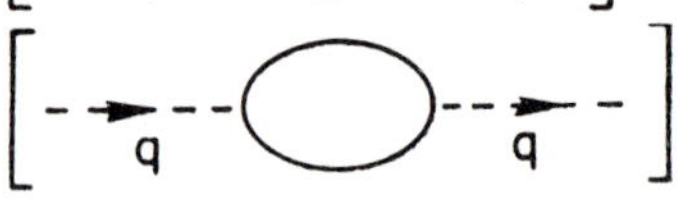

if we can just cast J in this form. Introducing the Fourier representations for $\rho_{ex}(\mathbf{r}t)$, the operator $\rho(\mathbf{r}t)$, etc., we find for J,

$$J = \frac{1}{\Omega}\int \frac{d^4q}{(2\pi)^4}\, \rho_{\rm ex}(\mathbf{q}, \nu)\rho_{\rm ex}(-\mathbf{q}, -\nu)\; v(\mathbf{q}) \int d\tau\, e^{i\nu\tau} \langle \Psi_0 | T(\rho(-\mathbf{q}, -\tau)\rho(\mathbf{q}0)) | \Psi_0 \rangle$$

which is just the desired form. Thus we obtain

$$= -\frac{(v(\mathbf{q}))^2}{\Omega}\int_{-\infty}^{\infty} d\tau\, e^{i\nu\tau} \langle \Psi_0 | T(\rho(-\mathbf{q}, -\tau)\rho(\mathbf{q}0)) | \Psi_0 \rangle,$$

and so

$$P_t(\mathbf{q}, \nu) = -i\Omega^{-1}\int_{-\infty}^{\infty} d\tau\, e^{i\nu\tau} \langle \Psi_0 | T(\rho(-\mathbf{q}, -\tau)\rho(\mathbf{q}0)) | \Psi_0 \rangle. \quad (3.81)$$

Just as we introduced the Lehmann representation for G (which was also defined by a T-product matrix element), we can show that

$$P_t(\mathbf{q}, \nu) = \int_{C_t} d\nu' \frac{\mathscr{F}(\mathbf{q}, \nu')}{\nu' - \nu}, \quad (3.82)$$

where $\mathscr{F}(\mathbf{q}, \nu)$ is as given in (3.79) and

C_t =

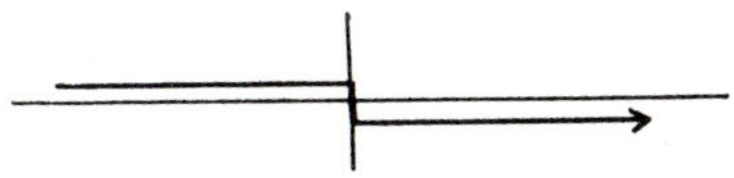

The following other properties are simple to show (particularly if one follows the development given for the analytic properties of $G(p)$):

(*a*) $P_t(\mathbf{q}, \nu) = P_t(\mathbf{q}, -\nu)$. (3.83a)

(*b*) $Im\, P_t(\mathbf{q}, \nu) = \pi \mathscr{F}(\mathbf{q}, \nu)$ for $\nu > 0$, real;
$= -\pi \mathscr{F}(\mathbf{q}, \nu)$ for $\nu < 0$, real. (3.83b)

(*c*) $$Re\, P_t(\mathbf{q}, \nu) = \frac{1}{\pi}\mathscr{P}\int_0^{\infty} d\nu' \frac{2\nu'}{\nu'^2 - \nu^2} Im\, P_t(\mathbf{q}, \nu')$$

and $$P_t(\mathbf{q}, \nu) = \frac{1}{\pi}\int_0^{\infty} d\nu' \frac{2\nu'}{(\nu' - i\eta)^2 - \nu^2} Im\, P_t(\mathbf{q}, \nu'). \quad (3.83c)$$

(d) $P_t(\mathbf{q}, \nu)$ goes into its complex conjugate as ν crosses the contour. (3.83d)

The time-ordered product and retarded commutator polarization propagators are very simply related to one another. For ν on the positive real axis, they are the same, since the contour C_t can then be deformed to C_r:

$$P_r(\mathbf{q}, \nu > 0) = P_t(\mathbf{q}, \nu > 0). \tag{3.84}$$

Once either propagator is known on the positive real axis, both propagators are then known, in principle, everywhere, by virtue of their respective analytic properties and symmetry properties [i.e., by (3.80c) or (3.83c)].

3. *Ground State Energy*

A simple proof of equation (3.65), useful in expressing the ground state energy in terms of the dielectric constant, can be given directly from the Lehmann representation. This and several other properties of the dielectric constant (or polarization propagator) are discussed in this way by Nozières and Pines.[7] To prove (3.65) consider, from (3.81),

$$\begin{aligned}\int_{-\infty}^{\infty} \frac{d\nu}{2\pi} P_t(\mathbf{q}, \nu) &= -i\,\Omega^{-1} \int d\tau \delta(\tau) \langle \Psi_0 | T(\rho(-\mathbf{q}, -\tau)\rho(\mathbf{q}0)) | \Psi_0 \rangle \\ &= -i\Omega^{-1} \langle \Psi_0 | \rho(-\mathbf{q})\rho(\mathbf{q}) | \Psi_0 \rangle. \end{aligned} \tag{3.85}$$

Therefore,

$$\begin{aligned}\int \frac{d^4q}{(2\pi)^4} P_t(q) v(\mathbf{q}) &= -i \left\langle \Psi_0 \,\middle|\, \Omega^{-1} \sum_q v(\mathbf{q})\rho(-\mathbf{q})\rho(\mathbf{q}) \,\middle|\, \Psi_0 \right\rangle \\ &= -2i\Omega^{-1} \langle \Psi_0 | H_1 | \Psi_0 \rangle. \end{aligned}$$

Since the right side is pure imaginary, we have

$$\langle \Psi_0 | H_1 | \Psi_0 \rangle = -\tfrac{1}{2}\Omega \int \frac{d^4q}{(2\pi)^4} Im\, P_t(\mathbf{q}) v(\mathbf{q}) = -\tfrac{1}{2}\Omega \int \frac{d^4q}{(2\pi)^4} Im \frac{1}{\kappa_t(q)}, \tag{3.86}$$

which is just (3.65). Since $\kappa_t(\mathbf{q}, \nu) = \kappa_t(\mathbf{q}, -\nu)$, we can write

$$\langle \Psi_0 | H_1 | \Psi_0 \rangle = -\Omega \int \frac{d^3q}{(2\pi)^3} \int_0^{\infty} \frac{d\nu}{2\pi} Im \frac{1}{\kappa_t(q)}.$$

Because $\kappa_t(\mathbf{q}, \nu) = \kappa_r(\mathbf{q}, \nu)$ for $\nu > 0$, we have

$$\langle\Psi_0|H_1|\Psi_0\rangle = -\Omega \int \frac{d^3q}{(2\pi)^3} \int_0^\infty \frac{d\nu}{2\pi} Im \frac{1}{\kappa_r(q)}, \tag{3.87}$$

which is the form of the relation derived by Nozières and Pines.

4. *Asymptotic Behavior for $P_r(\mathbf{q}, \nu)$*

From (3.78) we can get the asymptotic behavior of $P_r(\mathbf{q}, \nu)$ for large ν, fixed $\mathbf{q}$. The idea is that, for sufficiently large $\mathscr{E}_n - \mathscr{E}_0$ and fixed $\mathbf{q}$, the numerators like $|\langle n|\rho(q)|0\rangle|^2$ become arbitrarily small so we can neglect such terms. Then choose ν at least as large as these $\mathscr{E}_n - \mathscr{E}_0$, so that we can expand the denominators in $(\mathscr{E}_n - \mathscr{E}_0)/\nu$ obtaining

$$P_r(\mathbf{q}, \nu) \underset{\nu \to \infty}{\simeq} -\frac{1}{\Omega\nu} \sum_n \left\{ |\langle n|\rho(q)|0\rangle|^2 \left(1 - \frac{\mathscr{E}_n - \mathscr{E}_0}{\nu} + \ldots\right) - |\langle n|\rho(-q)|0\rangle|^2 \left(1 + \frac{\mathscr{E}_n - \mathscr{E}_0}{\nu} + \ldots\right)\right\}.$$

Assuming inversion symmetry, $|\langle n|\rho(q)|0\rangle|^2 = |\langle n|\rho(-q|0\rangle|^2$, we have

$$P_r(\mathbf{q}, \nu) \underset{\nu \to \infty}{\simeq} \frac{1}{\Omega\nu^2} \sum_n (\mathscr{E}_n - \mathscr{E}_0)\{\langle 0|\rho(-q)|n\rangle\langle n|\rho(q)|0\rangle + \langle 0|\rho(q)|n\rangle\langle n|\rho(-q)|0\rangle\}.$$

The summation on the right appears in a very general f-sum rule which we now derive.

The factors $(\mathscr{E}_n - \mathscr{E}_0)$ can be obtained by judicious insertion of the total Hamiltonian H in various places in the matrix elements as follows:

$$\begin{aligned} P_r(\mathbf{q}, \nu) &\underset{\nu \to \infty}{\simeq} \frac{1}{\Omega\nu^2} \sum_n \{\langle 0|\rho(-q)|n\rangle\langle n|[H, \rho(q)]|0\rangle - \langle 0|[H, \rho(q)]|n\rangle\langle n|\rho(-q)|0\rangle\} \\ &= \frac{1}{\Omega\nu^2}\langle 0|[\rho(-q),\ [H, \rho(q)]]|0\rangle. \end{aligned} \tag{3.88}$$

Now it happens that the double commutator is just a c-number which can be readily evaluated. It is somewhat easier to do in the notation of first quantization, in which

$$H = \sum_{i=1}^{N} \frac{p_i^2}{2m} + \frac{1}{2} \sum_{i \neq j} v(\mathbf{r}_i - \mathbf{r}_j),$$

and

$$\rho(\mathbf{q}) = \sum_i e^{-i\mathbf{q}\cdot\mathbf{r}_i}.$$

Then

$$[\mathrm{H}, \rho(\mathbf{q})] = \sum_{i,j} \left[\frac{p_i^2}{2m}, e^{i\mathbf{q}\cdot\mathbf{r}_j}\right] = \sum_i \left[\frac{p_i^2}{2m}, e^{-i\mathbf{q}\cdot\mathbf{r}_i}\right]$$

$$= -\frac{1}{2m} \sum_i (2\mathbf{p}_i \cdot \mathbf{q} + q^2)\, e^{-i\mathbf{q}\cdot\mathbf{r}_i};$$

$$[\rho(-\mathbf{q}), [\mathrm{H}, \rho(\mathbf{q})]] = \left[\sum_j e^{i\mathbf{q}\cdot\mathbf{r}_j}, \sum_i \frac{-1}{2m}(2\mathbf{p}_i \cdot \mathbf{q} + q^2)\, e^{-i\mathbf{q}\cdot\mathbf{r}_i}\right]$$

$$= -\sum_i \frac{\mathbf{q}}{m} \cdot [e^{i\mathbf{q}\cdot\mathbf{r}_j}, \mathbf{p}_i\, e^{-i\mathbf{q}\cdot\mathbf{r}_i}]$$

$$= \sum_i \frac{q^2}{m} = \frac{Nq^2}{m},$$

or

$$\sum_n (\mathscr{E}_n - \mathscr{E}_0) |\langle 0|\rho(\mathbf{q})|n\rangle|^2 = \frac{Nq^2}{2m}, \tag{3.89}$$

the general f-sum rule, which does not make use of the particular form of $v(\mathbf{q})$. Finally,

$$P_r(\mathbf{q}, \nu) \underset{\nu \to \infty}{\sim} \frac{q^2 N}{\nu^2 m \Omega}. \tag{3.90}$$

For a gas of electrons, for which $v(\mathbf{q}) = 4\pi e^2/q^2$,

$$v(\mathbf{q}) P_r(\mathbf{q}, \nu) \underset{\nu \to \infty}{\sim} \frac{\nu_{pl}^2}{\nu^2} \quad \text{and} \quad \kappa(\mathbf{q}, \nu) \underset{\nu \to \infty}{\sim} 1 - \frac{\nu_{pl}^2}{\nu^2}, \tag{3.90'}$$

where

$$\nu_{pl} = \left(\frac{4\pi Ne^2}{m\Omega}\right)^{\frac{1}{2}},$$

the classical plasma frequency in a gas of N/Ω electrons per unit volume.

5. *A Sum Rule for Im* $1/\kappa(q)$

The asymptotic behavior of $P_r(\mathbf{q}, \nu)$ allows us to deduce a sum rule involving the dielectric constant. Consider the following integral

$$J = \int_{-\infty}^{\infty} P_r(\mathbf{q}, \nu)\nu \, d\nu. \tag{3.91}$$

Because of (3.80a), $ReP_r(\mathbf{q}, \nu)$ is even in ν and contributes nothing:

$$J = i\int_{-\infty}^{\infty} Im\, P_r(\mathbf{q}, \nu)\nu \, d\nu = 2i\int_{0}^{\infty} Im\, P_r(\mathbf{q}, \nu)\nu \, d\nu.$$

$$= \frac{2i}{v(\mathbf{q})}\int_{0}^{\infty} Im\left(\frac{1}{\kappa_r(q)}\right)\nu \, d\nu. \tag{3.92}$$

We can also evaluate J by closing the contour in (3.91) by a large semicircle in the upper half-plane. Since $P_r(\mathbf{q}, \nu)$ has no poles in the upper half-plane, the integral along the entire contour is zero. Thus

$$J = \int_{-\infty}^{\infty} P_r(\mathbf{q}, \nu)\nu \, d\nu = -\int_{\curvearrowleft} P_r(\mathbf{q}, \nu)\nu \, d\nu.$$

But the integral along the large semicircle can be performed exactly knowing the asymptotic behavior (3.90):

$$\int_{\curvearrowleft} P_r(\mathbf{q}, \nu)\nu \, d\nu = \frac{i\pi q^2 N}{m\Omega}$$

Thus we have the sum rule

$$\int_{0}^{\infty} Im\frac{1}{\kappa_r(q)}\,\nu \, d\nu = -\frac{\pi \nu_{pl}^2}{2}, \tag{3.93}$$

a result which is exact. It can also be shown (see Du Bois[8], Appendix B) that

$$\int_{0}^{\infty} Im\kappa_r(q)\,\nu \, d\nu = \frac{\pi \nu_{pl}^2}{2}.$$

6. *The Stopping Power in Terms of Im*$(1/\kappa)$

Another quantity that can be expressed in terms of the dielectric constant is the stopping power for fast particles. Suppose we have a fast moving test charge (not necessarily a point charge):

$$Ze\,\rho_{\text{ex}}(\mathbf{r}-\mathbf{v}t)$$

where $Ze\,\rho_{\text{ex}}$ is the particle density in the test charge distribution This external charge induces a charge in the electron-hole medium which in turn produces an electric field

$$-\nabla_r \int \frac{e\rho_{\text{in}}(\mathbf{r}'t)}{|\mathbf{r}-\mathbf{r}'|} d^3r'$$

exerting a force on the test charge,

$$-Ze^2 \int\int \rho_{\text{ex}}(\mathbf{r}-\mathbf{v}t)\nabla_r \frac{\rho_{\text{in}}(\mathbf{r}'t)}{|\mathbf{r}-\mathbf{r}'|} d^3r\, d^3r'.$$

To keep the charge moving at constant velocity, an equal and opposite force from us is required, so that the rate of power input to the charge (and hence the medium) is

$$\begin{aligned}\left(\frac{dW}{dt}\right)_{t=0} &= Ze^2 \int \rho_{\text{ex}}(\mathbf{r}-\mathbf{v}t)\mathbf{v}\cdot\nabla_r \frac{\rho_{\text{in}}(\mathbf{r}'t)}{|\mathbf{r}-\mathbf{r}'|} d^3r\, d^3r' \\ &= -Ze^2 \int \frac{\mathbf{v}\cdot\nabla_r\rho_{\text{ex}}(\mathbf{r}-\mathbf{v}t)\rho_{\text{in}}(\mathbf{r}'0)}{|\mathbf{r}-\mathbf{r}'|} d^3r\, d^3r', \end{aligned} \tag{3.94}$$

Let us introduce Fourier transforms for everything so that we can relate the induced charge to the test charge via the dielectric constant. Noting that

$$\rho_{\text{ex}}(q) = \rho_{\text{ex}}(\mathbf{q})\cdot 2\pi\,\delta(\nu-\mathbf{q}\cdot\mathbf{v})$$

so that

$$\rho_{\text{in}}(\mathbf{r}, 0) = \int \frac{d^3q}{(2\pi)^3} \frac{\rho_{\text{ex}}(\mathbf{q})}{\kappa_r(\mathbf{q}, \mathbf{q}\cdot\mathbf{v})} \exp(i\mathbf{q}\cdot\mathbf{r}),$$

we find

$$\frac{dW}{dt} = i4\pi Ze^2 \int \frac{d^3q}{(2\pi)^3} \frac{\mathbf{v}\cdot\mathbf{q}}{q^2} \frac{|\rho_{\text{ex}}(\mathbf{q})|^2}{\kappa_r(\mathbf{q}, \mathbf{q}\cdot\mathbf{v})}.$$

Since this must be real

$$\frac{dW}{dt} = -4\pi Ze^2 \int \frac{d^3q}{(2\pi)^3} \frac{\mathbf{v} \cdot \mathbf{q}}{q^2} |\rho_{\mathrm{ex}}(\mathbf{q})|^2 \, Im\left(\frac{1}{\kappa_r(\mathbf{q}, \mathbf{q} \cdot \mathbf{v})}\right). \tag{3.95}$$

For a fast point charge, $\rho_{\mathrm{ex}}(\mathbf{q}) = 1$ and $v \to \infty$:

$$\begin{aligned}
\frac{dW}{dt} &= \frac{-4\pi Ze^2}{(2\pi)^3} \cdot 2\pi \int_0^\infty dq \int_{-1}^{1} vqx \, Im \frac{1}{\kappa_r(\mathbf{q}, qvx)} dx \\
&= -\frac{Ze^2}{\pi v} \int_0^\infty \frac{dq}{q} \int_{-vq}^{vq} \nu \, Im \frac{1}{\kappa_r(\mathbf{q}, \nu)} d\nu \\
&\underset{v \to \infty}{\sim} -\frac{2Ze^2}{\pi v} \int_0^\infty \frac{dq}{q} \int_0^\infty \nu \, Im \frac{1}{\kappa_r(\mathbf{q}, \nu)} d\nu,
\end{aligned} \tag{3.96}$$

a well-known expression for the stopping power.

For a further discussion of the dielectric constant, see Nozières and Pines,[7] Quinn and Ferrell,[9] and Du Bois[10] where still more references are given.

J. THE HIGH DENSITY ELECTRON GAS

Until now we have restricted ourselves to an analysis that was either exact or limited only by the validity of infinite order perturbation theory. Let's now consider an electron gas at high density, for which certain approximations, simple in the formalism we have developed, seem appropriate.

The reason we consider a high density gas is that this is equivalent to considering the limit of very weak Coulomb interactions so we expect some sort of perturbation theory to work. It is easy to see this on purely dimensional grounds.

Suppose that the volume per particle Ω/N corresponds to a sphere of radius r_0. Let us use r_0 as a basic unit of length. Then $\hbar/r_0$ is the unit of momentum. Writing the Hamiltonian in first quantization

$$H = \sum_i \frac{p_i^2}{2m} + \sum_{i,j}{}' \frac{e^2}{r_{ij}}$$

and letting $r' = r/r_0$ and $p' = (p/\hbar)r_0$, we obtain

$$H = \frac{\hbar^2}{mr_0^2}\left(\sum_i \frac{p_i'^2}{2} + \frac{me^2 r_0}{\hbar^2} \sum_{i,j}{}' \frac{1}{r'_{ij}}\right).$$

Recall that the Bohr radius is $a_0 = \hbar^2/me^2$. If we define the dimensionless parameter r_s as

$$r_s = \frac{r_0}{a_0} = \left(\frac{3\Omega}{4\pi N}\right)^{1/3} \frac{me^2}{\hbar^2}, \tag{3.97}$$

then we have

$$\frac{H}{me^4/\hbar^2} = \frac{1}{r_s^2}\left(\sum_i \frac{p_i'^2}{2} + r_s \sum_{i,j}{}' \frac{1}{r_{ij}'}\right), \tag{3.98}$$

where the right-hand side is now dimensionless. This shows that the kinetic energy goes as r_s^{-2}. The lowest order in r_s coming from the potential energy corresponds to the diagram

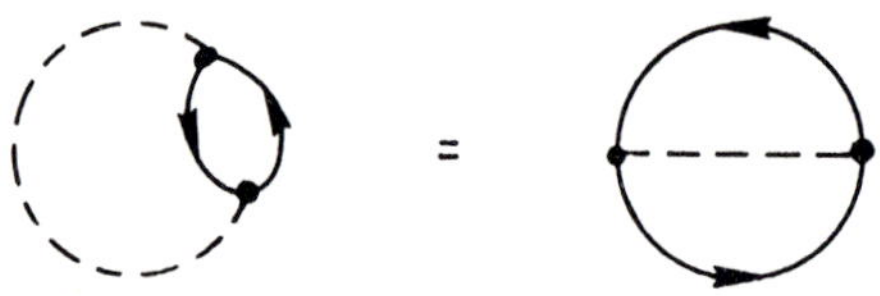

which, if you write it down you will see, is just the first-orde- exchange energy of the Fermi sea, and goes like $1/r_s$. All other contributions to the ground state energy go as r_s^0, $r_s \ln r_s$, or higher powers of r_s times powers of $\ln r_s$. Why they do not go just as higher powers of r_s will be evident shortly. High density is equivalent to small r_s. But by (3.97) it is also equivalent to small e^2. Thus high density and weak interaction are equivalent.

It might be handy to list a few simple dimensional relations in these units.

$$\frac{P_F}{\hbar} = \frac{1.92}{r_0} = \frac{1.92}{r_s a_0}, \quad \left(1.92 = \pi\left(\frac{3}{2\pi}\right)^{2/3}\right)$$

$$E_F = \tfrac{1}{2}\left[\pi\left(\frac{3}{2\pi}\right)^{2/3}\right]^2 \frac{E_0}{r_s^2} \simeq 1.85\frac{E_0}{r_s^2} \text{ where } E_0 = \frac{me^4}{\hbar^2}$$

$$\hbar\nu_{pl} = \frac{3^{1/2}E_0}{r_s^{3/2}}$$

$$\frac{E_F}{\hbar\nu_{pl}} = \frac{1.061}{r_s^{1/2}}$$

The theory of the free electron gas at high densities has been developed in several papers by several authors. Unfortunately, the purely high density theory seems to be valid at most for $r_s \lesssim 2$, whereas the values of r_s in the most favorable metals are $r_s \gtrsim 4$. Still, the high density theories give a certain amount of insight and also may be the point of departure for more accurate theories in the future. For extensions and interpolations to intermediate densities see Hubbard,[11] Nozières and Pines,[12] DuBois,[13] and Ferrell.[14]

The formal perturbation expansion and diagrammatic methods may be applied directly to the electron gas provided a minor modification is made to include the electron spin. Each particle line must now carry a spin variable s, which does not change in any interaction. When the value of s on any line is not fixed by the conditions on external lines, then s must be summed over its two possible values. Since the effects of the exclusion principle have already been included, these additional sums, when they occur, lead simply to factors of two.

1. *Evaluating the Polarization Propagator in the Pair Approximation*

Recall expressions (3.63) and (3.65) for the potential energy:

$$\langle \Psi_0 | H_1 | \Psi_0 \rangle = \frac{i\Omega}{2} \int \frac{d^4q}{(2\pi)^4} \mathscr{V}(q) \Lambda(q) = -\frac{\Omega}{2} \int \frac{d^4q}{(2\pi)^4} Im \frac{1}{\kappa_t(q)}.$$

Also recall expressions for the dielectric constant in terms of the contributions from proper polarization subdiagrams:

$$\kappa_t(q) = 1 - v(q)\Lambda(q) = \kappa_1 + i\,\kappa_2. \tag{3.56}$$

The simplest approximation for $K_t(q)$, the so-called pair approximation, is to approximate $\Lambda(q)$ in (3.56) by

$\Lambda(q) \cong$ 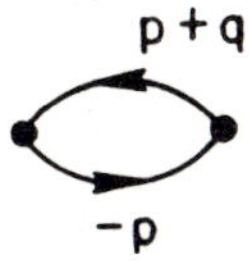

This is equivalent to approximating the diagrams (all of which are connected vacuum diagrams) contributing to (3.63) by

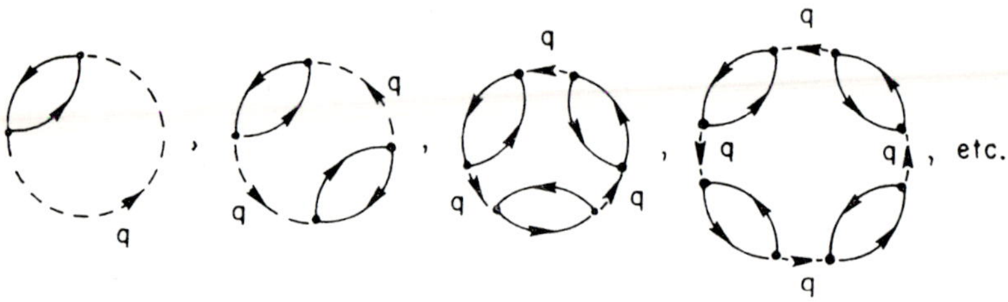

The suggestion to consider only such diagrams was first put forth and carried out by Gell-Mann and Brueckner.[15] The argument was that, of all possible vacuum diagrams of a particular order, the diagram of the above form would have the most factors of $1/q^2$ (in other diagrams of the same order, interaction lines appear with momenta like $\mathbf{q}+\mathbf{p}'-\mathbf{p}$) and therefore would lead to the most singular integrals at $\mathbf{q} = 0$. By summing the most singular diagrams of every order, before carrying out the integration, it was hoped to obtain a finite result, and the dominant contribution. The hope was borne out.

The important feature of the pair approximation is that whenever an electron and hole are present in a diagram, they always have the same total 4-momentum q as any other electron-hole pair in the same diagram, i.e., there is no mixing of electron-hole pairs of momentum q and q'. The possibility of treating the various Fourier components of the charge density fluctuation independently is the basis of the "random phase approximation" of Bohm and Pines[16] and also of the approach of Sawada,[17] Wentzel,[18] Sawada, Brueckner, Fukuda and Brout,[19] and of Brout.[20] All these approaches are nondiagrammatic ways of making more or less the same approximation.

Let us then calculate $\Lambda(q)$ in this approximation, assuming ν is real.

$$\Lambda(q) = -2i \int \frac{d^4p}{(2\pi)^4} \frac{1}{E+\nu-\epsilon(\mathbf{p}+\mathbf{q})+i\delta_{p+q}} \frac{1}{E-\epsilon(\mathbf{p})+i\delta_p}.$$

The factor of 2 is from the two possible spins of the electron which we are going to take into consideration. In doing the E-integral there are three possible cases:

(*a*) If both $\mathbf{p}$ and $\mathbf{p}+\mathbf{q}$ are on the same side of the Fermi surface, both poles are on the same side of the real axis. Closing the contour on the other side, we see that the integral gives zero.

(*b*) If $\mathbf{p}$ is below and $\mathbf{p}+\mathbf{q}$ is above the Fermi surface, we close the contour above getting a contribution from the pole at $E = \epsilon(\mathbf{p})-i\delta_p$:

$$\int \frac{d^3p}{(2\pi)^3} \frac{2f_p(1-f_{p+q})}{\nu+\epsilon(\mathbf{p})-\epsilon(\mathbf{p}+\mathbf{q})+i\delta}. \tag{3.100a}$$

The factor $f_p(1-f_{p+q})$, where $f_p = 1$ for $|\mathbf{p}| < p_F$ and $f_p = 0$ for $|\mathbf{p}| > p_F$, is introduced to avoid an explicit restriction on the region of the p-integration.

(*c*) If $\mathbf{p}$ is above and $\mathbf{p}+\mathbf{q}$ is below the Fermi surface, we close above getting a contribution from the pole at $E = \epsilon(\mathbf{p}+\mathbf{q})-\nu-i\,\delta_{p+q}$:

$$\int \frac{d^3p}{(2\pi)^3} \frac{2f_{p+q}(1-f_p)}{\epsilon(\mathbf{p}+\mathbf{q})-\nu-\epsilon(\mathbf{p})+i\delta}. \tag{3.100b'}$$

Letting $\mathbf{p} = \mathbf{p}+\mathbf{q}$ and $\mathbf{q}' = -\mathbf{q}$ and dropping primes in this expression, we effectively swap $\mathbf{p}$ and $\mathbf{p}+\mathbf{q}$ and obtain

$$\int \frac{d^3p}{(2\pi)^3} \frac{2f_p(1-f_{p+q})}{\epsilon(\mathbf{p})-\nu-\epsilon(\mathbf{p}+\mathbf{q})+i\delta}. \tag{3.100b}$$

$\Lambda(q)$ is the sum of (3.100a) and (3.100b). In the sum of the principal parts of these integrals, we can neglect f_{p+q} because

$$\int \frac{d^3p}{(2\pi)^3} \frac{f_p f_{p+q} \omega_{pq}}{\nu^2-\omega_{pq}{}^2}$$

changes sign under the swap $\mathbf{p} \rightleftarrows \mathbf{p}+\mathbf{q}$ ($\omega_{pq} \equiv \epsilon(\mathbf{p}+\mathbf{q})-\epsilon(\mathbf{p})$ is odd under the swap). This can be represented pictorially as follows.

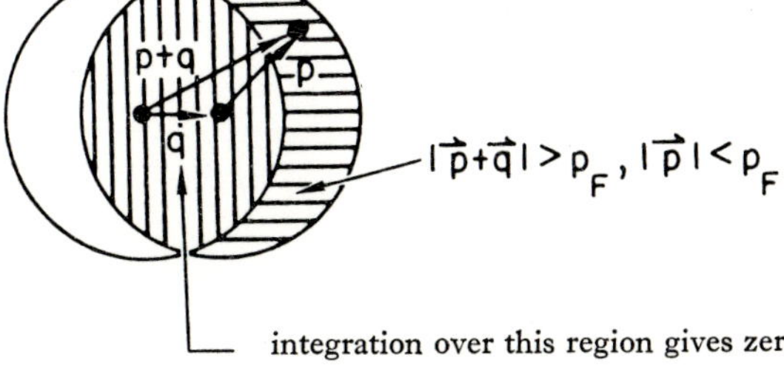

integration over this region gives zero

In the sum of the imaginary parts of (3.100a) and (3.100b) f_{p+q} can again be neglected, but it is more convenient for interpretation to retain this factor.

We finally obtain for $\kappa_t(q)$, when ν is real,

$$\begin{aligned}\kappa_t(q) &= 1-v(q)\,\mathscr{P}\int\frac{d^3p}{(2\pi)^3}f_p\,\frac{4\omega_{pq}}{\nu^2-\omega_{pq}{}^2}\\ &\qquad -2\pi i\,v(q)\int\frac{d^3p}{(2\pi)^3}\,f_p(1-f_{p+q})[\delta(\nu-\omega_{pq})+\delta(\nu+\omega_{pq})]\\ &\equiv \kappa_1(q)+i\kappa_2(q). \end{aligned}\tag{3.101}$$

Observe that $\Lambda(q)$, and hence $P_t(q)$, are even functions of ν, in agreement with (3.83a). *Henceforth we shall assume, without loss of generality, that* $\nu > 0$.

The integrals for $\kappa_1(q)$ and $\kappa_2(q)$ can be evaluated straightforwardly and lead to expressions first derived by Lindhard:[21]

$$\begin{aligned}\kappa_1(q) = 1+\frac{v(\mathbf{q})p_F}{2\pi^2}\Bigg\{1+\frac{1-(\nu'+\tfrac{1}{2}q')^2}{2q'}\ln\left|\frac{1+(\nu'+\tfrac{1}{2}q')}{1-(\nu'+\tfrac{1}{2}q')}\right|\\ -\frac{1-(\nu'-\tfrac{1}{2}q')^2}{2q'}\ln\left|\frac{1+(\nu'-\tfrac{1}{2}q')}{1-(\nu'-\tfrac{1}{2}q')}\right|\Bigg\}\end{aligned}\tag{3.102a}$$

and*

$$\kappa_2(q)=\begin{cases} 0 \text{ for } \nu > \dfrac{q^2+2qp_F}{2m} \\[2ex] 0 \text{ for } q > 2p_F \text{ and } \nu < \dfrac{q^2-2qp_F}{2m} \\[2ex] \dfrac{2e^2m^2\nu}{q^3} \text{ for } q < 2p_F \text{ and } 0 < \nu < -\left(\dfrac{q^2-2qp_F}{2m}\right) \\[2ex] \dfrac{e^2m^2p_F{}^2}{q^3}\{1-(\nu'-\tfrac{1}{2}q')^2\} \\ \qquad \text{for } 0 < -\left(\dfrac{q^2-2qp_F}{2m}\right) < \nu < \dfrac{q^2+2qp_F}{2m} \end{cases}\tag{3.102b}$$

where

$$q' = q/p_F \text{ and } \nu' = \nu/(qp_F/m).$$

* Here and in what follows, the symbol q will represent $|\mathbf{q}|$ as well as the four-vector $(\mathbf{q}, \nu)$. There should be no reason for confusion, however.

2. *Discussion of the Dielectric Constant and Polarization Propagator*

There are several features of the dielectric constant and polarization propagator worth noting. First, observe that κ_2, the imaginary part of the dielectric constant, is zero unless the frequency ν and the momentum $\mathbf{q}$ are so related that it is possible to excite an electron out of the Fermi sea (thus creating a hole) so that the electron-hole pair will have energy ν and momentum $\mathbf{q}$. It is clear that in the limit that ν tends to zero, the number of possible electron-hole pairs that can be created tends to zero, and thus the imaginary part of the dielectric constant (and hence of the polarization propagator) tends to zero. It is also clear that if q is either too small or too large, for fixed ν, it will not be possible to create a pair with energy ν.

We can represent these and more detailed properties of (3.102a, b) in Figure 1, where the nature of $\kappa_2(q)$ in the various regions of $(\mathbf{q}, \nu)$ space is indicated.

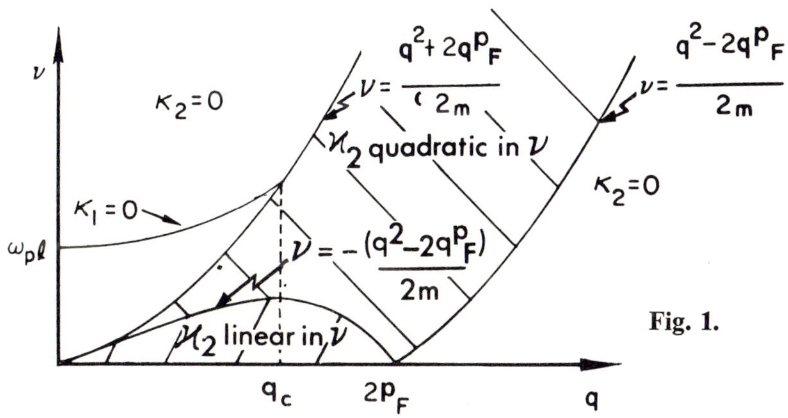

Fig. 1.

Second, consider the dielectric constant in the limit that $\nu \to 0$. As we have just mentioned, it is real. This limit is important because it gives the screening of static impurity charges, for which $w(\underline{q}) = w(q)\,\delta(\nu)$. It is also important in the interaction of electrons with vibrations of the positive ions in the solid, because these vibrations can be considered to have very low frequencies so long as the speed of sound is much smaller than the electron velocity at the Fermi surface. In this limiting case we find

$$\kappa_1(\mathbf{q},\ 0) = 1+0.66r_s\left(\frac{p_F}{q}\right)^2 u(q/2p_F) \qquad (3.103)$$

where

$$u(x) = \tfrac{1}{2}\left[1+\frac{1-x^2}{2x}\ln\left|\frac{x+1}{x-1}\right|\right].$$

We plot $1/\kappa_1$ roughly in Figure 2.

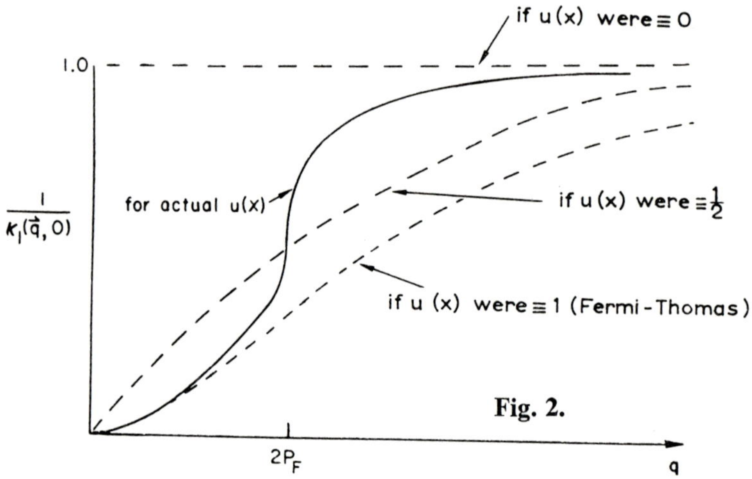

Fig. 2.

The important characteristics of $u(x)$ are the following:

(*a*) $u(0) = 1$ so that $\kappa_1(\mathbf{q}, 0) \underset{q\to 0}{\simeq} 1+0.66\, r_s\left(\frac{p_F}{q}\right)^2$

$$= \frac{q^2 + 3(\hbar v_{pl})^2 v^2{}_{\text{fermi}}}{q^2}$$

(*b*) $u(\infty) = 0$ so that $\kappa_1 \underset{q\to\infty}{\simeq} 1$

(*c*) $u(q = 2p_F) = \frac{1}{2}$

(*d*) $du/dq \to \infty$ as $q \to 2p_F$.

The small q behavior of $\kappa_1(\mathbf{q}, 0)$ means that at large distances the potential due to a point charge is completely screened out. In fact, if $\kappa_1(\mathbf{q}, 0)$ had its long wavelength form for *all* $\mathbf{q}$ (which it does not), a point charge would have a Yukawa potential going like

$$\int d^3q \frac{\exp(-i\mathbf{q}\cdot\mathbf{r})}{q^2} \frac{q^2}{q^2+\lambda^2} \sim e^{-\lambda r}/r, \qquad \lambda^2 = 0.66 r_s p_F{}^2.$$

This is essentially the Fermi-Thomas result. The deviation of $\kappa_1(\mathbf{q}, 0)$ from this simple behavior, and in particular the infinite derivative implied by characteristic (d) above (which in turn is a result of the sharpness of the Fermi surface*), yield instead an induced charge density, potential, electric field, and electric field gradient each having as its leading term an expression of the form

$$\frac{\cos(2p_F r+\phi)}{r^3}$$

where ϕ is an appropriate phase for each quantity. The actual potential with this surprisingly long range and oscillating behavior was first derived and recognized by Friedel,[22] and discussed in detail from the point of view of the many-body perturbation theory by Langer and Vosko.[23] Recent experiments by T. J. Rowland[24] seem to confirm this long range behavior, according to detailed calculations by Kohn and Vosko.[25] Rowland found a marked decrease in the nuclear magnetic resonance absorption of copper when a wide variety of metallic atoms are introduced substitutionally in small concentrations. The only explanation consistent with these and other experiments, is that the electric field gradients set up by the solute atoms interact significantly with the quadrupole moments of as many as 85 of the neighboring copper nuclei.

A third important feature of the dielectric constant is that for small enough q, there exists a real frequency ν which makes not only

*The sharpness of the Fermi surface, within the pair approximation, has been investigated by E. Daniel and S. H. Vosko, *Phys. Rev.* **120,** 2041 (1960).

$\kappa_2(q)$ but also $\kappa_1(q)$ vanish, i.e., there exists a real pole of the polarization propagator. To see this quantitatively we need only let q become small in (3.101), which has been written in just such a form as to make this limit obvious. We see that

$$\kappa_1(q) \simeq 1 - \frac{8\pi e^2}{m\nu^2} \int \frac{d^3p}{(2\pi)^3} f_p = 1 - \frac{4\pi e^2 N}{m\nu^2 \Omega} = 1 - \frac{\nu_{pl}{}^2}{\nu^2}, \qquad (3.104)$$

so that the pole of the polarization propagator is at ν_{pl}, the classical electron plasma frequency. As q increases, it can be shown that this pole shifts slightly to higher frequencies as indicated in Figure 1.

The fact that the polarization propagator has a pole on the real axis for small q indicates that (in the pair approximation) there is an elementary excitation of this energy which is undamped. These are the well-known plasmons. So long as we limit ourselves to the pair approximation, the only kind of damping possible for the plasmons is through the decay into an electron-hole pair. This is only possible if there exist electron-hole pairs of the same momentum and energy as the plasmon, which is possible only in the region where $\kappa_2(q) \neq 0$. In fact as we go to shorter wavelengths (higher q), the plasmon curve in Figure 1 enters this region. Suddenly the decay rate of the plasmons rises, so rapidly in fact that they cease to be well-defined elementary excitations.

The lack of any damping of the plasmons for small q is a consequence of the pair approximation. Actually, the pole of the polarization propagator will always have an imaginary part whenever there is an intermediate state in a polarization propagator diagram having the same energy as the plasmon. The pair approximation restricts us to intermediate states with just one pair present. More complicated diagrams would allow two or more pairs present simultaneously, and these would allow the plasmon to decay for all wave vectors $\mathbf{q}$. These are higher order diagrams, however, so that at high densities these processes are relatively weak. A thorough discussion of this point is given by Du Bois,[26] who considers diagrams of the following kind for the polarization propagator.

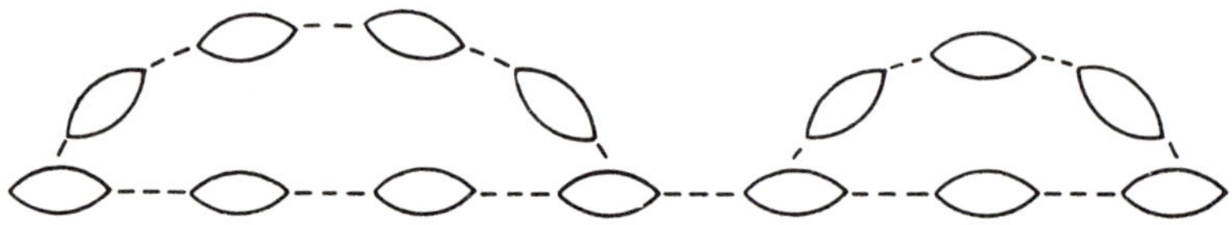

Inclusion of such diagrams not only modifies the imaginary part of the pole of the polarization propagator but it also modifies the real part, i.e., it changes the plasmon self-energy somewhat. The shift and broadening of the plasmon energies may have an effect on the characteristic energy losses noted when fast electrons pass through thin metallic films—losses which are primarily due to plasmons, for small momentum transfers by the incident electron. These points are all discussed by Du Bois.[27]

The existence of the plasmons shows itself in another interesting way. In the expression for the ground state energy, we must evaluate

$$\tfrac{1}{2}\int\frac{d^3q}{(2\pi)^3}\int\frac{d\nu}{2\pi}\frac{\kappa_2}{\kappa_1{}^2+\kappa_2{}^2}.$$

Now in higher approximations, as we have remarked, κ_2 is never zero. It is however small at the plasmon energy. We approximate the integrand by its limit as $\kappa_2 \to 0$, which is just $\pi\delta(\kappa_1)$. The ν-integration then gives a contribution from the plasmon pole. When this is evaluated, and the integration over the electron charge is performed, we find for each plasmon mode a contribution to the ground-state energy of $\hbar\nu_{pl}/2$, the zero-point energy of the plasmon (in the long wavelength approximation).

Du Bois[28] has shown how to rewrite the polarization propagator to bring out even more clearly the existence of the plasmon as an elementary excitation. Recall the dispersion relation (3.83c) for the polarization propagator, which can be written

$$P_t(\mathbf{q},\nu) = \frac{1}{\pi v(\mathbf{q})}\int_{+} d\nu' \frac{2\nu'}{\nu'^2-\nu^2}\frac{Im\,\kappa(\mathbf{q},\nu')}{|\kappa(\mathbf{q},\nu')|^2}. \tag{3.105}$$

In (3.101) we found $Im\kappa(q)$, but only for ν real. In (3.105) we need $Im\kappa(\mathbf{q},\nu)$ for ν just below the positive real axis. Even when $Im\kappa(\mathbf{q},\nu)$ vanishes for ν real, the displacement $\nu \to \nu - i\eta$ introduces an infinitesimal imaginary part, which must be included in (3.105). Assuming $\eta < \delta$ and ν positive, and recalling that $\kappa = 1 - v\Lambda$, we must now consider

$$\Lambda(\mathbf{q},\nu-i\eta)$$
$$= 2\int\frac{d^3p}{(2\pi)^3}f_p(1-f_{p+q})\left\{\frac{1}{(\nu-i\eta)-\omega_{pq}+i\delta}-\frac{1}{(\nu-i\eta)+\omega_{pq}-i\delta}\right\}$$

Then for those ($\mathbf{q}$, ν) such that $\nu = \omega_{pq}$ for some $\mathbf{q}$-momentum pair excitation (i.e. for some $|\mathbf{p}| < p_F$ and $|\mathbf{p}+\mathbf{q}| > p_F$), we get a finite imaginary part just as before; but if $\nu \neq \omega_{pq}$ for all $\mathbf{q}$-pair excitations, we get

$$Im\Lambda(\mathbf{q},\, \nu-i\eta) = -2\int \frac{d^3p}{(2\pi)^3} f_p(1-f_{p+q})\left\{\frac{\delta-\eta}{(\nu-\omega_{pq})^2}+\frac{\delta+\eta}{(\nu+\omega_{pq})^2}\right\}.$$

Since both integrals in the imaginary part are positive, we have a net negative imaginary part for $\Lambda(\mathbf{q},\, \nu-i\eta)$. We can thus write

$$\Lambda(\mathbf{q},\, \nu-i\eta) \underset{\nu > 0}{=} 2\mathscr{P}\int \frac{d^3p}{(2\pi)^3} f_p(1-f_{p+q})\frac{2\omega_{pq}}{\nu^2-\omega_{pq}}$$
$$-2\pi i\left[\int \frac{d^3p}{(2\pi)^3}\delta(\nu-\omega_{pq})f_p(1-f_{p+q}) + \Delta(\mathbf{q},\, \nu)\right]. \quad (3.106)$$

$\Delta(\mathbf{q},\, \nu)$ is zero or a positive infinitesimal, depending on whether or not there exists a $\mathbf{q}$-pair excitation of energy ν.

When (3.106) is inserted into (3.105) the polarization propagator naturally consists of two terms:

$$P_t(\mathbf{q},\, \nu) = 2\int_0^\infty d\nu' \frac{2\nu'}{(\nu'-i\eta)^2-\nu^2}\frac{1}{|\kappa(\mathbf{q},\, \nu'-i\eta)|^2}$$
$$\times\left[\int \frac{d^3p}{(2\pi)^3}\delta(\nu'-\omega_{pq})f_p(1-f_{p+q}) + \Delta(\mathbf{q},\, \nu)\right],$$
$$= P_1+P_2.$$

For P_1 one finds simply

$$P_1 = 2\int \frac{d^3p}{(2\pi)^3} f_p(1-f_{p+q})\frac{2\omega_{pq}}{(\omega_{pq}-i\eta)^2-\nu^2}\cdot\frac{1}{|\kappa(\mathbf{q},\, \omega_{pq}-i\eta)|^2}.$$

This is the sum of propagators for all electron-hole pairs of momentum $\mathbf{q}$, including a coupling factor $|\kappa(\mathbf{q},\, \omega_{pq})|^{-2}$. For P_2, because $\kappa_2(\mathbf{q},\, \nu')$ is infinitesimal (or zero), one can simplify the ν'-integration by recalling that

$$\frac{\kappa_2}{\kappa_1{}^2+\kappa_2{}^2} \underset{\kappa_2 \to 0}{\longrightarrow} \pi\delta(\kappa_1).$$

Because the ν'-integration is restricted to the region where there are no $\mathbf{q}$-pair excitations, the only solution to $\kappa_1(\mathbf{q}, \nu) = 0$ is $\nu_{pl}(\mathbf{q})$, and this solution exists only if $q < q_c$, the wave vector at which the plasmon spectrum merges with the pair excitation continuum. Thus

$$P_2(\mathbf{q}, \nu) = \frac{2}{v(\mathbf{q})} \frac{2\pi\nu_{pl}}{(\nu_{pl}-i\eta)^2-\nu^2} \frac{1}{(d\kappa_1/d\nu)_{\nu_{pl}}} \text{ if } q < q_c,$$
$$= 0 \text{ if } q > q_c.$$

This is the plasmon propagator, and includes a coupling factor

$$2\pi\left[v(q)\left(\frac{d\kappa_1}{d\nu}\right)_{\nu_{pl}}\right]^{-1}$$

An expression like

$$\frac{\omega_q}{(\omega_q-i\eta)^2-\nu^2}$$

has the form of a propagator for a boson of energy ω_q, as we shall see in the next chapter, when we discuss phonons.

Although this decomposition of $P_t(q)$ brings out the boson-like role played by plasmons and electron-hole pairs, it is of course not so convenient a form for calculation as is obtained directly from (3.101) and $v(\mathbf{q})\, P(\mathbf{q}) = [\kappa(\mathbf{q})]^{-1} - 1$.

3. *Electron Green Function*

Corrections to the electron Green function can also be calculated if we approximate the self-energy part $\Sigma(p)$ by the sum of all diagrams represented by

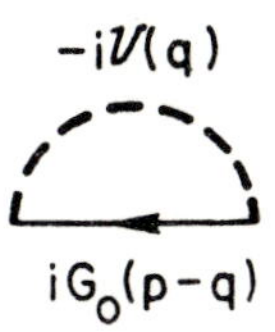

For the high density gas, one further makes the pair approximation for $\mathscr{V}(q)$. The detailed calculations are given by Du Bois[29] and by Quinn and Ferrell.[30] Let me just summarize some of the results.

First, the single particle energies are modified due to the cloud of virtual electron-hole pairs and plasmons that each carries with it.

This changes the density of states of the elementary excitations near the Fermi surface, which in turn affects the specific heat, reducing it according to Du Bois for $r_s \lesssim 0.8$ and enhancing it for actual metals. The singular behavior in the density of states that one obtains if one includes in $\Sigma(p)$ only the first-order exchange corrections, represented by the diagram

but does not include further correlation corrections disappears in this more accurate treatment. This has been discussed in detail by Quinn and Ferrell.

The $(p-p_F)^2$ dependence of the imaginary part of the self-energy near the Fermi surface has been proved to all orders of perturbation theory by Luttinger[31] (assuming the validity of the perturbation expansion and re-summation approach) and is the basis of his theoretical justification of the Landau theory of Fermi liquids. Physically this means that (neglecting scattering from phonons or imperfections) elementary excitations near the Fermi surface are long-lived, allowing us to speak of their real energies and to do the statistical analysis leading to the usual expression for the specific heat. The imaginary part of the energy gives the rate of decay of these excitations due to emission of single electron-hole pairs (i.e., excitation of an electron out of the Fermi sea) and single plasmons. For fast excitations, this imaginary part becomes appreciable, so the excitations cease to be long-lived. In fact, as we have seen, the stopping power can be related to this imaginary part.

4. *Ground State Energy*

The evaluation of the ground state energy using the pair approximation follows from (3.59) and (3.65). Thus

$$\mathscr{E} = \mathscr{E}_0 - \int_0^{e^2} \frac{d(e'^2)}{e'^2} \left[\tfrac{1}{2} N v(\mathbf{r}=0;\ e'^2) + \tfrac{1}{2} \int \frac{d^4 q}{(2\pi)^4} Im \frac{1}{\kappa(q)} \right].$$

In the pair approximation, we see that $Im[1/\kappa(q)]$ has the form

$$Im\frac{1}{\kappa} = \frac{-\kappa_2}{\kappa_1^2+\kappa_2^2} = \frac{-e'^2k_2}{(1+e'^2k_1)^2+e'^2k_2^2}$$

where k_1 and k_2 are independent of e'^2. The integrations on e'^2 can be performed immediately, so that

$$\mathscr{E} = \mathscr{E}_0+\tfrac{1}{2}\int\frac{d^3q}{(2\pi)^3}\left[\int_{-\infty}^{\infty}\frac{d\nu}{2\pi}\tan^{-1}\frac{\kappa_2(\mathbf{q},\nu)}{\kappa_1(\mathbf{q},\nu)}-\frac{4\pi Ne^2}{q^2}\right].$$

When the further integrations are performed, as discussed by Hubbard,[32] one obtains for the energy per particle the result first given by Gell-Mann and Brueckner:

$$\frac{\mathscr{E}}{N} = \frac{e^2}{2a_0}\left[\frac{2.21}{r_s^2}-\frac{0.916}{r_s}+0.0622\ln r_s-0.096+\ldots\right]$$

The first two terms are the kinetic energy and exchange energy of the Fermi sea, and the remaining terms contribute to the so-called correlation energy.

REFERENCES

1. For the most general treatment at finite temperatures, see P. C. Martin and J. Schwinger, *Phys. Rev.* **115**, 1342, section IV.
2. V. M. Galitskii and A. B. Migdal, *Soviet Phys. JETP* **7**, 96 (1958).
3. M. Gell-Mann and F. Low, *Phys. Rev.* **84**, 350 (1951) Appendix.
4. W. Kohn and J. M. Luttinger, *Phys. Rev.* **118,** 41 (1960).
5. J. M. Luttinger, and J. C. Ward, *Phys. Rev.* **118,** 1417 (1960).
6. V. M. Galitskii, *Soviet Phys. JETP* **7**, 104 (1958).
7. P. Nozières and D. Pines, *Nuovo Cimento* **9**, 470 (1958).
8. D. F. DuBois, *Ann. Phys.* **8**, 24 (1959).
9. J. J. Quinn and R. A. Ferrell, *Phys. Rev.* **112**, 612 (1958).
10. D. F. DuBois, *Ann. Phys.* **7**, 174 (1959).
11. J. Hubbard, *Proc. Roy. Soc.* (*London*) **A240**, 539 (1957); **A243**, 336 (1957).
12. P. Nozières and D. Pines, *Phys. Rev.* **111**, 442 (1958), and other papers.
13. D. F. DuBois, *Ann. Phys.* **8**, 24 (1959).
14. R. A. Ferrell, *Phys. Rev. Letters*, **1**, 443 (1958).
15. M. Gell-Mann and K. A. Brueckner, *Phys. Rev.* **106**, 364 (1957).

16. D. Pines and D. Bohm, *Phys. Rev.* **85**, 338 (1952); **92**, 608 (1953); and D. Pines, *Rev. Mod. Phys.* **28**, 184 (1956). Also, several papers in *Phys. Rev.* **109** and **113** by P. Nozières and D. Pines.
17. K. Sawada, *Phys. Rev.* **106**, 372 (1957).
18. W. Wentzel, *Phys. Rev.* **108**, 1593 (1957).
19. K. Sawada, K. A. Brueckner, N. Fukuda, and R. Brout, *Phys. Rev.* **108**, 507 (1957).
20. R. Brout, *Phys. Rev.* **108**, 515 (1957).
21. J. Lindhard, *Kgl. Danske Videnskap. Selskab. Mat.-fys. Medd.* **28**, No. 8, 1954.
22. J. Friedel, *Phil. Mag.* **43**, 153, 1952; *Advances in Physics*, **3**, 446, 1954; *Suppl. Nuovo Cimento*, **2**, 287, 1958.
23. J. Langer and S. H. Vosko, *J. Phys. Chem. Solids*, **12**, 196, 1959.
24. T. J. Rowland, *Phys. Rev.*, **119**, 900, 1960.
25. W. Kohn and S. H. Vosko, *Phys. Rev.* **119**, 912 (1960).
26. D. F. DuBois, *Ann. Phys.* **8**, 24 (1959), Section II.
27. *Ibid.*, Section III.D.
28. D. F. Du Bois, *Annals of Physics*, **7**, 174, 1959, Section IV.
29. D. F. Du Bois, *Ann. Phys.* **8**, 24 (1959), Section IIIE and IV.
30. J. J. Quinn and R. A. Ferrell, *Phys. Rev.* **112**, 612 (1958).
31. J. M. Luttinger, *Phys. Rev.* **121**, 942 (1961).
32. J. Hubbard, *Proc. Roy. Soc.* (*London*) **A243**, 336 (1957).

CHAPTER IV

THE ELECTRON-PHONON SYSTEM AT ABSOLUTE ZERO

A. General Formulation

Let us now include the positive ions in the metal, not as a smeared-out uniform charge distribution, but as a dynamical system coupled to the electrons. It is possible to formulate this problem as one of the interactions of two quantized fields. I shall discuss some of the formal features of this problem and the perturbation theory approach, just as I did for the electron system alone. In fact, much of the discussion involves only simple generalizations of what has been said already.

We are now considering a system of N electrons and N monovalent ions (for simplicity) in a volume Ω with a Hamiltonian

$$H = \sum_i \frac{p_i^2}{2m} + \frac{1}{2}\sum_{ij}{}' v(\mathbf{r}_i - \mathbf{r}_j) + \sum_\alpha \frac{P_\alpha^2}{2M} + \frac{1}{2}\sum_{\alpha\beta} V(\mathbf{R}_\alpha - \mathbf{R}_\beta) - \sum_{i\alpha} u(\mathbf{r}_i - \mathbf{R}_\alpha). \tag{4.1}$$

Here $\mathbf{p}_i$ and $\mathbf{r}_i$ refer to the ith electron; $\mathbf{P}_\alpha$ and $\mathbf{R}_\alpha$, to the αth ion. Only the conduction electrons are being treated separately. All other electrons are assumed to follow the motions of the ions perfectly. This is a reasonable assumption if the gap between the conduction band and lower bands is at all appreciable.

At first it might seem natural to consider H as decomposing into the three rows of (4.1), an electron Hamiltonian $H_{\rm el}$, an ion (ultimately "phonon") Hamiltonian H_{ph}, and an electron-ion interaction Hamiltonian H_{int}. But as was suggested for the electron gas alone, it is more convenient to include in H_{el} one-half the interaction of the electrons with the $\mathbf{q} = 0$ Fourier component of the ionic charge density (i.e. with the field of a uniform distribution of ionic charge). Similarly, in H_{ph} we include one-half the interaction of the ions with the $\mathbf{q} = 0$ Fourier component of the electronic charge density. These two terms must then be explicitly excluded from the interaction term H_{int}. Thus if we express the three interaction terms in (4.1) by means of the Fourier components of the electron and ion number densities,

$$\frac{1}{2}\sum_{\mathbf{q}} v(\mathbf{q})\rho_-(\mathbf{q})\rho_-(-\mathbf{q}) + \frac{1}{2}\sum_{\mathbf{q}} V(\mathbf{q})\rho_+(\mathbf{q})\rho_+(-\mathbf{q}) - \sum_{\mathbf{q}} u(\mathbf{q})\rho_+(\mathbf{q})\rho_-(-\mathbf{q}),$$

and if we make use of the obvious equalities

$$\rho_-(\mathbf{q} \to 0) = \rho_+(\mathbf{q} \to 0) = \frac{N}{\Omega}$$

and

$$v(\mathbf{q} \to 0) = V(\mathbf{q} \to 0) = u(\mathbf{q} \to 0),$$

then we may subtract $\frac{1}{2}v(0)\rho_-(0)\rho_+(0) = \frac{1}{2}v(0)\rho_\pm(0)\rho_\pm(0)$ from the first two terms and add twice it to the third, eliminating the $\mathbf{q} = 0$ terms from all three sums.

Now H_{el}, so modified, has been analyzed in the previous chapter by perturbation theoretic methods, assuming weak interactions between electrons. The formal structure of H_{ph} is exactly the same, but because of the much larger mass of the ions, the effective expansion parameter r_s is increased by a factor M/m and so the approach used for electrons fails completely. In fact, just as one hopes to consider the electrons a dense or weakly interacting charged gas, one must treat the ions as a dilute or strongly interacting many-body system. Such systems have been studied by several authors,[1] some with the electron system in mind in the hope that a study of both the strong and weak coupling limits might give insight into the actual intermediate coupling situation occurring in metals. In the strong coupling limit, however, the detailed nature of the charges (whether they are fermions or bosons, whether they are point charges or somewhat extended) plays an unimportant role. We may use the findings of these authors to write H_{ph} in the most advantageous way, i.e., by introducing normal modes.

Consider then a system of N monovalent ions of mass M, in the limit that $M \to \infty$. The ions interact in pairs through a potential $V(\mathbf{R}_\alpha - \mathbf{R}_\beta)$ which at large separations will be Coulombic but which may include a van der Waals term and some short-range repulsion. The ions also interact with a compensating uniform background of negative charge. In the zeroth approximation, only the potential energy plays a role; the ions are then perfectly localizable and arrange themselves to minimize this energy. Any Bravais lattice arrangement, in which each ion is at a center of symmetry, will be metastable and can be the point of departure for higher approximations. It may be interesting to note that the body-centered cubic, followed by the face-centered cubic, hexagonal close-packed, and simple cubic structures, appear to have the lowest energy for a given density, although the differences are very slight.

In the next approximation, the ions are allowed to vibrate and we must expand the potential term in H_{ph} to second order in the displacements from whichever metastable or stable configuration resembles the solid under investigation. The full potential is

$$\begin{aligned} U(\mathbf{R}_1 \ldots \mathbf{R}_N) &= \frac{1}{2} \sum_{\alpha\beta}' V(\mathbf{R}_\alpha - \mathbf{R}_\beta) - \sum_\alpha \int d^3r \frac{N}{\Omega} u(\mathbf{R}_\alpha - \mathbf{r}) \\ &= U(\mathbf{R}_1{}^0 \ldots \mathbf{R}_N{}^0) + \frac{1}{2} \sum_{\alpha\beta} \sum_{\mu\nu} \left(\frac{\partial^2 U}{\partial R_{\alpha\mu} \partial R_{\beta\nu}} \right)_0 \delta R_{\alpha\mu} \delta R_{\beta\nu}. \end{aligned} \tag{4.2}$$

Henceforth we neglect the constant $U(\mathbf{R}_1{}^0 \ldots \mathbf{R}_N{}^0)$.

First, a few qualitative remarks may be useful. If the uniform background density (and hence the ionic density) is low enough, the ions do not overlap. For rigid ions the interactions among ions and the interaction between each ion and the background charge in all Wigner-Seitz cells but its own can then all be taken to be Coulombic. Only the interaction of each ion with the background in its own Wigner-Seitz cell is then not rigorously Coulombic, but to simplify the discussion, we shall assume it to be, too. Thus we assume in H_{ph} that all potentials v, V and u are Coulombic. The diagonal elements of the $3N$-dimensional matrix $(\partial^2 U/\partial R_{\alpha\mu} \partial R_{\beta\nu})_0$ are readily understood. If the αth ion is given a displacement $\delta\mathbf{R}_\alpha$, it experiences forces from two sources: a force from the background charge in its own unit cell and a force from the charge distribution of all other cells. The force from its own cell is that from a uniform spherical distribution of the same volume, plus small corrections (which we neglect) from the corners included or omitted in replacing the WS cell by a sphere. It can be obtained from Poisson's equation for the electrostatic potential Φ:

$$\nabla^2\Phi = \frac{4\pi Ne}{\Omega}$$

which has the solution

$$\Phi = \frac{4\pi Ne}{\Omega}\frac{R^2}{6}.$$

The electrostatic potentials arising from the other unit cells are all small (and produce fields that tend to cancel), because they result from higher multipoles in the charge distribution within each unit cell. Thus if only the diagonal elements of the force constant matrix were to be considered, each ion would behave like a three-dimensional harmonic oscillator with frequency

$$\omega = \left(\frac{\omega_{pl}^2}{3}\right)^{1/2} = \left(\frac{1}{3}\frac{4\pi Ne^2}{m\Omega}\right)^{1/2}.$$

The off-diagonal terms in the force constant matrix are like dipole-dipole interactions and have the effect of splitting this $3N$-fold degeneracy, producing three bands. In the two "transverse" bands $\omega(\mathbf{k}) \to 0$ as $\mathbf{k} \to 0$ while in the "longitudinal" band $\omega(\mathbf{k}) \to \omega_{pl}$ as $\mathbf{k} \to 0$. It will turn out that only the longitudinal band makes a significant contribution to H_{int}. Plots of ω vs. $\mathbf{k}$ for special directions in $\mathbf{k}$-space are shown by Clark (*op. cit.*) from detailed calculations within the point-ion approximation. Values of $\omega(\mathbf{k})$ for numerous points throughout the Brillouin zone have been calculated by Carr (*op. cit.*).

The systematic analysis of the Hamiltonian H_{ph} with the quadratic potential of (4.2) begins with an expansion of $\delta\mathbf{R}_\alpha$ in a set of plane waves, designed to reduce the force constant matrix from order $3N$ to N matrices of order 3:

$$\delta \mathbf{R}_\alpha = \frac{i}{N^{1/2}} \sum_{k\sigma} \bar{\phi}_{k\sigma} \boldsymbol{\epsilon}_{k\sigma} \exp(i\mathbf{k} \cdot \mathbf{R}_\alpha{}^0) \tag{4.3a}$$

Here $\mathbf{k}$ ranges over a set of quasi-continuous values (in the first Brillouin zone) determined by the periodic boundary conditions. The vectors $\boldsymbol{\epsilon}_{k\sigma}$, $\sigma = 1, 2, 3$, are a set of three orthonormal vectors to be chosen later for each $\mathbf{k}$. Without loss of generality we assume that $\boldsymbol{\epsilon}_{-k\sigma} = -\boldsymbol{\epsilon}_{k\sigma}$.

Similar expansions can be made for the ion momentum operators:

$$\mathbf{P}_\alpha = \frac{-i}{N^{1/2}} \sum_{k\sigma} \bar{\pi}_{k\sigma} \boldsymbol{\epsilon}_{k\sigma} \exp(-i\mathbf{k} \cdot \mathbf{R}_\alpha{}^0) \tag{4.3b}$$

The operators $\bar{\phi}_{k\sigma}$ and $\bar{\pi}_{k\sigma}$ can be obtained by inverting (4.3a, b):

$$\bar{\phi}_{k\sigma} = \frac{-i}{N^{1/2}} \sum_{\alpha} e^{-i\mathbf{k} \cdot \mathbf{R}_\alpha{}^0} (\boldsymbol{\epsilon}_{k\sigma} \cdot \delta \mathbf{R}_\alpha) \tag{4.4a}$$

$$\bar{\pi}_{k\sigma} = \frac{i}{N^{1/2}} \sum_{\alpha} e^{i\mathbf{k} \cdot \mathbf{R}_\alpha{}^0} (\boldsymbol{\epsilon}_{k\sigma} \cdot \mathbf{P}_\alpha) \tag{4.4b}$$

From the canonical commutation rules for the ion coordinates and momenta,

$$[\delta R_{\alpha\mu}, \delta P_{\beta\nu}] = i\delta_{\alpha\beta}\delta_{\mu\nu} \text{ and } [\delta \mathbf{R}_\alpha, \delta \mathbf{R}_\beta] = [\mathbf{P}_\alpha, \mathbf{P}_\beta] = 0,$$

we obtain

$$[\bar{\phi}_{k\sigma}, \bar{\pi}_{k'\sigma'}] = i\delta_{kk'}\delta_{\sigma\sigma'},$$
$$[\bar{\phi}_{k\sigma}, \bar{\phi}_{k'\sigma'}] = [\bar{\pi}_{k\sigma}, \bar{\pi}_{k'\sigma'}] = 0. \tag{4.5}$$

Also, from the hermiticity of $\delta \mathbf{R}_\alpha$ and $\mathbf{P}_\alpha$, we readily find that

$$\bar{\phi}_{k\sigma}{}^\dagger = \bar{\phi}_{-k\sigma}, \tag{4.6a}$$

$$\bar{\pi}_{k\sigma}{}^\dagger = \bar{\pi}_{-k\sigma}. \tag{4.6b}$$

H_{ph} can now be written in terms of these new operators:

$$H_{\text{ph}} = \sum_{k\sigma} \frac{\bar{\pi}_{k\sigma}\bar{\pi}_{-k\sigma}}{2M} + \frac{1}{2} \sum K_{k\sigma} \bar{\phi}_{k\sigma} \bar{\phi}_{-k\sigma} \tag{4.7}$$

where

$$K_{k\sigma} = \boldsymbol{\epsilon}_{k\sigma} \cdot \mathbf{f}(\mathbf{k}) \cdot \boldsymbol{\epsilon}_{k\sigma}$$

and

$$(\mathbf{f}(\mathbf{k}))_{\mu\nu} = \sum_{(R_\alpha{}^0-R_\beta{}^0)} \left(\frac{\partial^2 U}{\partial R_{\alpha\mu}\, R_{\beta\nu}}\right)_0 e^{i\mathbf{k}\cdot(\mathbf{R}_\alpha{}^0-\mathbf{R}_\beta{}^0)} .$$

The $\boldsymbol{\epsilon}_{k\sigma}$ are chosen to be the eigenvectors of the matrix $(\mathbf{f}(\mathbf{k}))_{\mu\nu}$. For point ions, the dynamical matrix $(\mathbf{f}(\mathbf{k}))_{\mu\nu}$ has been exhibited by Rosemary Coldwell-Horsfall and A. Maradudin and studied in the limit $\mathbf{k} \to 0$ by Maradudin and Weiss. The result, previously mentioned, is that for the longitudinal branch $\omega^2 \to \omega_{pl}{}^2$ and for the transverse branches $\omega^2 \to 0$. To obtain this result, it is important that the crystal is assumed to be a sphere of radius R, and that R tends to infinity before $\mathbf{k}$ tends to zero.

Now consider H_{int}, which when modified is just

$$H_{\text{int}} = -\sum_i \left[\sum_\alpha u(\mathbf{r}_i-\mathbf{R}_\alpha) - \frac{N}{\Omega}\int d^3R\, u(\mathbf{r}_i-\mathbf{R})\right].$$

If we expand this around $\{\mathbf{R}_\alpha{}^0\}$ and introduce the normal mode expansion (4.3a) we obtain, to first order,

$$H_{\text{int}} = \sum_i w(\mathbf{r}_i) - \frac{i}{N^{\frac{1}{2}}} \sum_{k\sigma,\, i} \phi_{k\sigma}\boldsymbol{\epsilon}_{k\sigma} \cdot \sum_\alpha [\nabla u(\mathbf{r})]_{r_i-\mathbf{R}_\alpha{}^0}\, e^{i\mathbf{k}\cdot\mathbf{R}_\alpha{}^0}, \tag{4.8}$$

where

$$w(\mathbf{r}_i) = \sum_\alpha u(\mathbf{r}_i-\mathbf{R}_\alpha{}^0) - \frac{N}{\Omega}\int d^3R\, u(\mathbf{r}_i-\mathbf{R}).$$

The term $\Sigma w(\mathbf{r}_i)$ is just the interaction of the electrons with the ions in their equilibrium configuration compensated by a uniform negative background. Since it now refers only to the electrons we include it in H_{el}. The second term is the actual electron-phonon interaction.

If we write the electron part of the Hamiltonian in second quantization, and include spin through the subscripts s and s', the entire Hamiltonian, with all the regrouping of terms, is then

$$H_{\text{el}} = H + H_{\text{ph}} + H_{\text{int}}, \tag{4.9}$$

where

$$H_{\text{el}} = \sum_{ps}\left(\epsilon_0(\mathbf{p}) - \frac{N}{\Omega}v(\mathbf{r}=0)\right)C_{ps}{}^\dagger C_{ps} + \frac{1}{\Omega}\sum_{ps}\sum_{q\neq 0} w(\mathbf{q})C_{p+qs}{}^\dagger C_{ps}$$
$$+ \frac{1}{2\Omega}\sum_{psp's'}\sum_{q\neq 0} v(\mathbf{q})C_{ps}{}^\dagger C_{p-qs}C_{p's'}{}^\dagger C_{p'+q's'},$$

$$H_{\text{ph}} = \frac{1}{2}\sum_{k\neq 0,\sigma}\left(\frac{\bar{\pi}_{k\sigma}\bar{\pi}_{-k\sigma}}{2M} + \frac{K_{k\sigma}}{2}\bar{\phi}_{k\sigma}\bar{\phi}_{-k\sigma}\right),$$

$$H_{\text{int}} = \frac{1}{\Omega}\sum_{ps\sigma}\sum_{k\neq 0}\sum_{k'\neq 0} \bar{g}_{kk'}\bar{\phi}_{k\sigma}C_{p+k's}{}^{\dagger}\, C_{ps}\ ,$$

and

$$\bar{g}_{kk'} = \int d^3r\ e^{-i(\mathbf{p}+\mathbf{k}')\cdot\mathbf{r}}$$

$$\times\ \left[-iN^{-\frac{1}{2}}\boldsymbol{\epsilon}_{k\sigma}\cdot\sum_{\alpha}\nabla u(\mathbf{r}-\mathbf{R}_{\alpha}{}^{0})\exp(i\mathbf{k}\cdot\mathbf{R}_{\alpha}{}^{0})\right]e^{i\mathbf{p}\cdot\mathbf{r}},$$

$$= N^{-\frac{1}{2}}\sum_{\alpha} e^{i(\mathbf{k}-\mathbf{k}')\cdot\mathbf{R}_{\alpha}{}^{0}}\ \boldsymbol{\epsilon}_{k\sigma}\cdot\mathbf{k}'u(\mathbf{k}'),$$

$$= N^{-\frac{1}{2}}\sum_{K}\delta_{k,\,k'+K}\ \boldsymbol{\epsilon}_{k\sigma}\cdot(\mathbf{k}-\mathbf{K})\ u(\mathbf{k}-\mathbf{K}).$$

The **K**'s are the vectors of the lattice in **k**-space. We shall neglect all contributions to the electron-lattice interaction except from $\mathbf{K} = 0$, a common idealization of the lattice in which it is treated as a vibrating continuum so that the reciprocal lattice degenerates to $\mathbf{K} = 0$. Thus $\bar{g}_{kk'} \to \bar{g}_k\,\delta_{kk'}$ and $\bar{g}_k = N^{\frac{1}{2}}kV(k)$. We shall also assume $\boldsymbol{\epsilon}_{k1}$ is in the direction of **k** (a strictly longitudinal mode) and $\boldsymbol{\epsilon}_{k2}$ and $\boldsymbol{\epsilon}_{k3}$ are perpendicular to **k** (strictly transverse).* Thus, the electrons interact only with the longitudinal modes, so we shall neglect the transverse modes.

Notice that we have written H in terms of C_p and $C_p{}^{\dagger}$ which refer to plane wave states, implying that $\sum_{pq} w(\mathbf{q})C_{p+q}{}^{\dagger}C_p$ is to be treated by perturbation theory. Actually, in many solids as I have mentioned, it is not wise to try to treat all of $\sum_{pq} w(\mathbf{q})C_{p+q}{}^{\dagger}C_p$ by perturbation theory. Rather, some effective periodic potential should be included in the unperturbed part of H_{el}, and subtracted from the interaction part of H_{el}. It might then be more appropriate

* For real crystals, this assumption is valid, if at all, only for certain directions of high symmetry; it is nevertheless approximately valid in many crystals for long wavelengths and greatly simplifies the formalism.

to work not in a plane wave representation, but in a Bloch wave representation. In this case the terms involving $w(\mathbf{q})$, $v(\mathbf{q})$ and $\bar{g}_k$ would have a more complicated structure.

It is convenient to make a scale transformation introducing new variables ϕ_k and π_k and a new coupling constant g_k by the transformations

$$\begin{aligned}\bar{\phi}_k &= \frac{\phi_k}{\sqrt{M\omega_0(\mathbf{k})}}\\ \bar{\pi}_k &= \pi_k\sqrt{M\omega_0(\mathbf{k})}\\ \bar{g}_k &= g_k\sqrt{M\omega_0(\mathbf{k})},\end{aligned} \tag{4.10}$$

where

$$\omega_0(\mathbf{k}) = \sqrt{\frac{K_k}{M}}.$$

Then, neglecting spin,

$$\begin{aligned}H = \sum_p \left[\epsilon_0(\mathbf{p}) - \frac{N}{\Omega}v(\mathbf{r}=0)\right]C_p{}^\dagger C_p + \frac{1}{\Omega}\sum_p \sum_{q\neq 0} w(\mathbf{q})C_{p+q}{}^\dagger C_p\\ + \frac{1}{2\Omega}\sum_{pp'}\sum_{q\neq 0} v(\mathbf{q})C_p{}^\dagger C_{p-q}C_{p'}{}^\dagger C_{p'+q}\\ + \frac{1}{2}\sum_{k\neq 0}\omega_0(\mathbf{k})(\pi_k\pi_{-k}+\phi_k\phi_{-k})\\ + \frac{1}{\Omega}\sum_p\sum_{k\neq 0} g_k\phi_k C_{p+k}{}^\dagger C_p.\end{aligned} \tag{4.11}$$

To get a feeling for the electron-ion problem, let us assume that the ions, in the absence of the conduction electrons, look like point charges. Then

$$V(\mathbf{k}) = -\frac{4\pi e^2}{k^2} = -v(\mathbf{k}). \tag{4.12a}$$

Also, as we have remarked,

$$\omega_0(\mathbf{k}) \underset{k\to 0}{=} \omega_{pl} \equiv \sqrt{\frac{4\pi\, e^2 N}{M}}, \tag{4.12b}$$

the classical ion plasma frequency. From (4.12a, b) we have

$$g_k = -\left(\frac{v(\mathbf{k})}{\omega_0(\mathbf{k})}\right)^{\frac{1}{2}} \omega_{pl} \underset{k\to 0}{\rightarrow} -(v(\mathbf{k})\omega_{pl})^{\frac{1}{2}}. \tag{4.13}$$

To compute matrix elements involving the ϕ_k and π_k operators, it is convenient to introduce creation and annihilation operators $b_k{}^\dagger$ and b_k defined as follows:

$$\phi_k = \frac{b_k + b_{-k}{}^\dagger}{\sqrt{2}} \quad \text{and} \quad \pi_k = \frac{b_{-k} - b_k{}^\dagger}{i\sqrt{2}}. \tag{4.14a}$$

The inverse relations are

$$b_k = \frac{\phi_k + i\pi_{-k}}{\sqrt{2}} \quad \text{and} \quad b_k{}^\dagger = \frac{\phi_{-k} - i\pi_k}{\sqrt{2}}. \tag{4.14b}$$

The commutation rules (4.5) for ϕ_k and π_k imply

$$[b_k, b_{k'}{}^\dagger] = \delta_{kk'}$$

and

$$[b_k, b_{k'}] = [b_k{}^\dagger, b_{k'}{}^\dagger] = 0. \tag{4.15}$$

The pure phonon part of the Hamiltonian is

$$H_{\text{ph}} = \sum \omega_0(\mathbf{k})(b_k{}^\dagger b_k + \tfrac{1}{2}). \tag{4.16}$$

Let me just recall some well-known facts for completeness. If we neglect the interactions with the conduction electrons, the elementary excitations of the many-ion system described by the Hamiltonian H_{ph} are quantum excitations of the harmonic motions of the amplitudes of the normal modes, with energies $\omega_0(\mathbf{k})$. The operator product $b_k{}^\dagger\, b_k$ is a number operator, telling the degree of excitation of the **k**th mode. We may associate with these elementary excitations discrete amounts of "field momentum", **k**. The total field momentum is then

$$\mathscr{P} = \sum \mathbf{k} b_k{}^\dagger b_k.$$

This operator commutes with the total pure phonon part of the Hamiltonian. It is called a "momentum" because it generates displacements of deformations in the lattice, i.e., $\exp(i\mathscr{P}\cdot\mathbf{R})$ applied to any quantum mechanical state of the lattice produces a new state physically the same as the first but describing motions displaced in the crystal through **R**.

The elementary excitations, having definite energies and momenta, can be thought of as particles, which are called phonons. The interaction term in the Hamiltonian thus describes the scattering of an electron from **p** to **p**+**k** with the creation of a phonon −**k** or the annihilation of a phonon **k**. The operator ϕ_k decreases the total phonon field momentum by **k**. If we had not made the continuum idealization, the interaction term would only conserve total momentum to within a vector of the lattice in **k**-space, which somewhat complicates things.

B. Green Functions Defined

Green functions can be introduced for the phonons in very close analogy with those for the electron-hole system. Without going into the calculation details, let me just list various one-phonon Green functions and show them explicitly for noninteracting phonons.

The basic phonon Green function for systems that are not necessarily homogeneous or time-independent is

$$D(\mathbf{k}t, \mathbf{k}'t') = i\langle\Psi_0|T(\phi_k(t)\phi_{-k'}(t'))|\Psi_o\rangle \tag{4.17}$$

where $|\Psi_0\rangle$ is the true ground state of the entire system and here the T-product is the same as the so-called P-product, i.e., no factors of (-1) are introduced when the operators are reordered because the basic operators satisfy commutation rules (phonons are thus bosons) rather than anti-commutation rules.

For time-independent systems, $D(\mathbf{k}t, \mathbf{k}'t')$ is replaced by

$$D(\mathbf{k}, \mathbf{k}', t) = i\langle\Psi_0|T(\phi_k(t)\phi_{-k'}(0))|\Psi_0\rangle; \tag{4.18}$$

and if, furthermore, the system is homogeneous, we need only

$$D(\mathbf{k},t) = i\langle\Psi_0|T(\phi_k(t)\phi_{-k}(0))|\Psi_0\rangle. \tag{4.19a}$$

Various Fourier transforms may then be defined in the usual way:

$$D(k) \equiv D(\mathbf{k}, \omega) = \int \frac{dt}{2\pi} \exp(i\omega t)\, D(\mathbf{k}, t), \tag{4.19b}$$

$$D(x) = \int \frac{d^4k}{(2\pi)^4} \exp(ik \cdot x)\; D(k), \tag{4.19c}$$

etc.,

where $$k = (\mathbf{k},\ \omega)$$

and $$k \cdot x = \mathbf{k} \cdot \mathbf{r} - \omega t\ .$$

Note that because H does not commute either with the phonon number operator $\Sigma_k\, b_k{}^\dagger b_k$ or the total phonon momentum operator $\Sigma_k\, \mathbf{k} b_k{}^\dagger b_k$, neither of these quantities is conserved and the ground state is not sharp in either of them. Within the continuum approximation, however, the *total* momentum $\Sigma_p \mathbf{p} C_p{}^\dagger C_p + \Sigma_k\, \mathbf{k} b_k{}^\dagger b_k$ does commute with H and is therefore sharply defined in the ground state. It is easy to show that the operator $\phi_k(t)$ decreases this total momentum by $\mathbf{k}$ for all t, so that within the continuum approximation $\langle \Psi_0 | \phi_k(t) \phi_{-k'}(0) | \Psi_0 \rangle = 0$ if $\mathbf{k} \neq \mathbf{k}'$.

For free phonons, the various Green functions are readily calculated, the calculations being simplified through the use of b_k and $b_k{}^\dagger$:

$$D_0(\mathbf{k},\ t) = \frac{i}{2}\, e^{-i\omega_0(\mathbf{k})|t|} \tag{4.20a}$$

and

$$\begin{aligned} D_0(k) &= \frac{1}{2}\left(\frac{1}{\omega + \omega_0(\mathbf{k}) - i\eta} - \frac{1}{\omega - \omega_0(\mathbf{k}) + i\eta} \right) \\ &= \frac{-\,\omega_0(\mathbf{k})}{\omega^2 - (\omega_0(\mathbf{k}) - i\eta)^2}. \end{aligned} \tag{4.20b}$$

We notice that

(1) $D_0(\mathbf{k},\ t)$ is even in t,
(2) $D_0(k)$ is even in ω,
(3) $D_0(k)$ has poles as a function of ω at $\pm(\omega_0(\mathbf{k}) - i\eta)$.

C. Exact Phonon Green Functions

Just as for the electron-hole system, we can obtain a spectral representation for the phonon Green functions by expanding in a complete set of exact intermediate states. We obtain, if we assume inversion symmetry for our system,

$$D(\mathbf{k},\ t) = i \int_0^\infty d\omega'\, \Phi(\mathbf{k},\ \omega') \exp(-i\omega'|t|) \tag{4.21a}$$

and

$$D(\mathbf{k},\,\omega) = \int_0^\infty d\omega'\,\Phi(\mathbf{k},\,\omega')\left(\frac{1}{\omega+\omega'-i\eta}-\frac{1}{\omega-\omega'+i\eta}\right), \quad (4.21b)$$

where

$$\Phi(\mathbf{k},\,\omega)d\omega = \sum_{\omega<\mathscr{E}_n-\mathscr{E}_0<\omega+d\omega} |(\phi_k)_{n0}|^2. \quad (4.22)$$

If we define

$$\begin{aligned}\mathscr{C}(\mathbf{k},\,\omega) &= \Phi(\mathbf{k},\,\omega),\ \omega > 0\\ &= -\Phi(\mathbf{k},\,-\omega),\ \omega < 0,\end{aligned} \quad (4.23)$$

then (4.21b) can be written

$$D(\mathbf{k},\,\omega) = \int \frac{\mathscr{C}(\mathbf{k},\,\omega')}{\omega'-\omega}d\omega'. \quad (4.24)$$

For noninteracting phonons

$$\Phi(\mathbf{k},\,\omega) = \tfrac{1}{2}\delta(\omega-\omega_0(\mathbf{k})).$$

From (4.21b) or (4.24) it is easy to show the usual properties arising from T-products:

(1) $Im\,D(\mathbf{k},\,\omega) = \pi\Phi(\mathbf{k},\,|\omega|)$, (4.25a)

(2) $$Re\,D(\mathbf{k},\,\omega) = -\frac{\mathscr{P}}{\pi}\int_0^\infty d\omega'\,Im\,D(\mathbf{k},\,\omega')\frac{2\omega'}{\omega^2-\omega'^2}, \quad (4.25b)$$

(3) $D(\mathbf{k},\,\omega)$ goes into $D^*(\mathbf{k},\,\omega)$ if ω just crosses over the contour. (4.25c)

The asymptotic properties of $D(\mathbf{k},\,t)$ for large t are governed by the poles of the analytic continuation of $D(\mathbf{k},\,\omega)$ for $\omega > 0$ into the lower half-plane. Note that the analytic continuation of $D(\mathbf{k},\,\omega)$ for $\omega < 0$ into the upper half-plane gives nothing new because

$$D(\mathbf{k},\,\omega) = D(\mathbf{k},\,-\omega). \quad (4.25d)$$

D. Perturbation Theory and Diagram

The perturbation theory we have developed for the many electron system goes through with only minor additions if we include the phonon system. Let us lump H_{ph} into the noninteracting part of the Hamiltonian, H_0, and include the electron-lattice term in the perturbation.

In the S-matrix development, we must now enlarge our definition of normal products. We now agree that in any normal product, all $b^\dagger$'s should stand to the left of b's and no factors of (-1) are introduced when the b's and $b^\dagger$'s are reordered. The Wick theorem now requires that we also sum over all possible contraction pairings of the ϕ_k's.

These additional requirements are most simply expressed in terms of modified rules for drawing momentum-space diagrams and calculating their contributions. We list the modifications:

(1) Pairs of points may now be connected by either a wavy line $\sim\!\sim\!\sim$ or a dotted line. Such a wavy line represents a contraction of phonon variables.

(2) Phonon lines (like Coulomb interaction lines) need not be directed, because $D(k) = D(-k)$. It is usually convenient to direct phonon lines just to keep everything consistent in a diagram.

(3) Each phonon line carries a factor $-iD_0(k)$ in evaluating the contribution of a diagram.

(4) Each vertex involving a phonon line carries a factor $-ig_k$ ($-i$ from perturbation theory). Since every interior line thus carries $-i(-ig_k)^2D_0(k)$, we might just as well have defined the Green function differently to absorb these factors. That is, we might define D in terms of $\phi'_k = ig_k\phi_k$ and this is in fact often done.[2]

(5) The counting is unchanged by the introduction of phonons, although now some lines are phonon lines and some lines are Coulomb lines and although n phonon lines come from $2n$th order perturbation theory while n Coulomb lines come from nth order perturbation theory.

E. Partial Summations of Diagrams

Partial summations of diagrams can still be effected although the presence of three kinds of lines and two kinds of vertices somewhat complicates things.

The exact one-particle Green function satisfies a Dyson equation

or

$$G = \left[G_0^{-1} - \Sigma \right]^{-1}, \tag{4.26}$$

where now, $-i\Sigma$ is the sum over all particle self-energy subdiagrams that cannot be split in two by cutting only one solid line, i.e. *proper* self-energy subdiagrams. Thus $-i\Sigma$ now includes

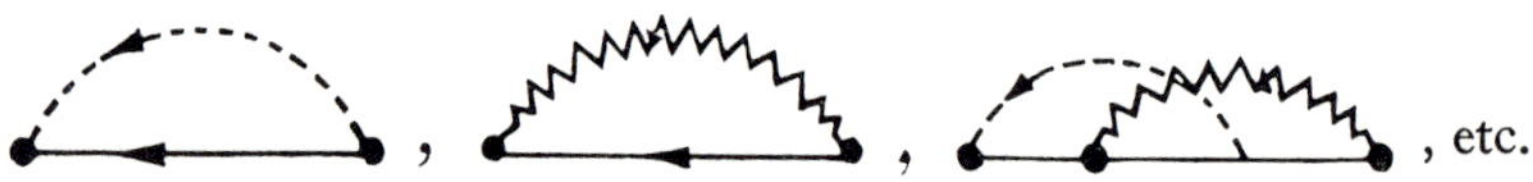

and is in general complex. One sums a large set of diagrams of any class when one considers only the restricted set of diagrams of that class that contain no particle self-energy subdiagrams (the so-called particle self-energy irreducible diagrams), but uses iG for each particle line.

A similar elimination of self-energy subdiagrams in phonon lines is more complicated because of the possibility of diagrams like the following:

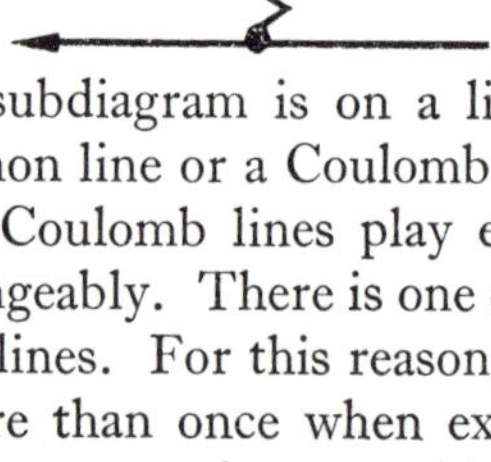

Here the self-energy subdiagram is on a line that is not, strictly speaking, either a phonon line or a Coulomb line. In most ways, in fact, the phonon and Coulomb lines play exactly equivalent roles and can enter interchangeably. There is one exception; only phonon lines are ever external lines. For this reason, and to avoid counting the same diagram more than once when executing various partial summations, we shall not treat these two kinds of lines in the same way.

To discuss this situation in detail, we must define two kinds of polarization (or, equivalently, phonon self-energy) subdiagrams. We first define $i\Lambda$ to be the sum over all polarization subdiagrams that cannot be split in two cutting *either* one phonon or one Coulomb line. Thus $i\Lambda$ includes

One can now define a generalized Coulomb line, $-i\mathscr{V}$, in analogy with the previous chapter, by

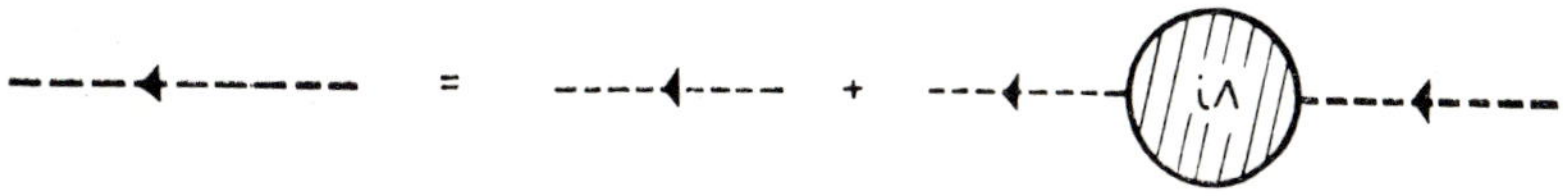

or

$$\mathscr{V} = \frac{v}{1-v\Lambda} \equiv \frac{v}{\kappa}. \tag{4.27}$$

It is important to emphasize, that though this quantity is analogous to $-i\mathscr{V}$ of the previous chapter, and reduces to it when $g \to 0$, it is not the complete interaction between two charges when $g \neq 0$. Thus $\kappa \equiv 1-v\Lambda$ is not the true dielectric constant of the electron-phonon system. Now define $i\Pi$ to be the sum over all polarization subdiagrams that cannot be split in two cutting one phonon line. $i\Pi$ includes, in addition to the diagrams of $i\Lambda$, diagrams like

It is easy to see that

$$\Pi = \frac{\Lambda}{1-v\Lambda} = \frac{1}{v}\left(\frac{1}{\kappa}-1\right). \tag{4.28}$$

The set of all diagrams contributing to the exact one-phonon Green function can be partially summed as follows:

so that

$$D = \frac{1}{D_0^{-1} - g^2\Pi} = \frac{D_0(1-v\Lambda)}{1-(v+g^2D_0)\Lambda}. \tag{4.29}$$

Now consider the situation of a diagram with an external phonon line entering (or leaving) the diagram with no polarization subdiagrams. A whole set of diagrams can be summed, in which the rest of the diagram is unchanged but the phonon line carries any number of polarization subdiagrams connected, in sequence, either by a phonon or Coulomb line. The summation can be represented as follows:

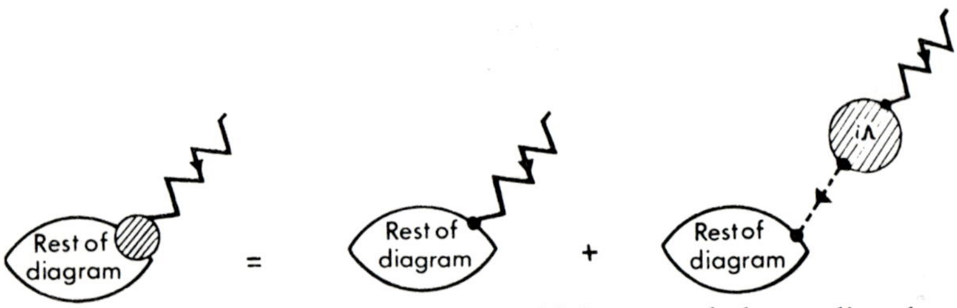

Thus we may draw only diagrams in which external phonon lines have no polarization subdiagrams, but to the phonon lines we must ascribe the contribution iD, and to the vertex where it joins the diagram we must ascribe

$$-i\Gamma = -i(g+g\mathscr{V}\Lambda),$$

$$\Gamma = \frac{g}{1-v\Lambda} = \frac{g}{\kappa}. \tag{4.30}$$

We shall see that, at least for small q, $\Gamma << g$.

Now consider a general interaction line between two points within a diagram. It will in general consist of either (or both) phonon and Coulomb lines, intercepted by proper polarization subdiagrams. We can, at first, divide all such lines into two classes: those in which only Coulomb lines occur between polarization subdiagrams, and those in which at least one phonon line occurs connecting two polarization subdiagrams. The sum over all diagrams of the first class yields just $-i\mathscr{V}$ for the interaction line. The sum over all diagrams of the second class yields the interaction

$$-iM = -i\Gamma^2 D = -i\frac{g^2D_0}{1-v\Lambda}\frac{1}{1-(v+g^2D_0)\Lambda} \tag{4.31}$$

represented by 

One can thus sum a large set of diagrams if one considers only diagrams that are irreducible in the sense of having no polarization parts, distinguishing however between the interactions

 and  .

and ascribing to them respectively $-i\mathscr{V}$ and $-iM$. Since whenever one of these interactions occurs, the other can also occur, we can draw instead one generalized interaction line

 ,

and ascribe to it the contribution

$$-i\mathscr{M} = -i(\mathscr{V}+M) = -i\frac{v+g^2D_0}{1-(v+g^2D_0)\Lambda}. \tag{4.32}$$

Further reductions are possible, but we shall not go into them here. What we have shown in these sections is that the system of bare ions and electrons interacting among themselves can be reduced to a system described by a considerably simpler set of diagrams, possessing the following restrictions:

(a) Phonon lines carry no self-energy subdiagrams, but represent a renormalized phonon Green function D (which is capable of describing decaying phonons).

(b) Particle lines carry no self-energy parts but represent a renormalized particle Green function G (which is capable of describing decaying particles).

(c) Coulomb lines carry no polarization subdiagrams but represent a screened Coulomb interaction $\mathscr{V}$.

(d) Phonons are never coupled to particle lines via Coulomb interactions, but instead of g an effective coupling Γ is introduced which, at least for long wavelengths, is drastically weaker.

(e) In addition to the screened Coulomb interaction between particles, there is also an effective interaction M between particles arising from the phonons. Among other complications, both interactions are non-instantaneous.

An interesting discussion of the electron-phonon system in the absence of Coulomb interactions is given along similar lines by Migdal.*

* *Ibid.*

Let us now consider some of the partial summations of this section in more detail.

F. Effect of Electron-Hole System on Phonons

The renormalized phonon of wave vector **k** has frequency $\omega(\mathbf{k})$ given by the pole of $D(\mathbf{k}, \omega)$. Setting the denominator of (4.29) equal to zero, and using (4.28) for Π, we have an equation for the renormalized frequency:

$$(\omega(\mathbf{k}))^2 = (\omega_0(\mathbf{k}))^2 + \frac{\omega_0(\mathbf{k}) g_k{}^2}{v(\mathbf{k})}\left(\frac{1}{\kappa(\mathbf{k}, \omega(\mathbf{k}))} - 1\right). \qquad (4.33)$$

In the point-ion approximation, at long wavelengths (4.33) becomes (recalling (4.12) and (4.13))

$$(\omega(\mathbf{k}))^2 = \frac{\omega_{pl}{}^2}{\kappa(\mathbf{k}, \omega(\mathbf{k}))} \simeq \frac{\omega_{pl}{}^2}{\kappa(\mathbf{k}, 0)}. \qquad (4.33')$$

Also, in this approximation

$$D(k) \simeq -\sqrt{\kappa(\mathbf{k}, 0)}\, \frac{\omega(\mathbf{k})}{\omega^2 - (\omega(\mathbf{k}) - i\eta)^2}. \qquad (4.34)$$

Aside from the factor $\sqrt{\kappa(k)}$ in (4.34), the so-called renormalization of the field, $D(k)$ in (4.34) describes a phonon propagating with frequency $\omega(\mathbf{k})$.*

* The reason for the extra factor $\sqrt{\kappa(\mathbf{k}, 0)}$ in $D(k)$ is easy to see. If the effective phonon frequency is $\omega(\mathbf{k})$ rather than $\omega_0(\mathbf{k})$, then the appropriate field to introduce in (4.10) is

$$\phi'_k = \bar{\phi}_k \sqrt{M\,\omega(\mathbf{k})} = \phi_k \left[\frac{\omega(\mathbf{k})}{\omega_0(\mathbf{k})}\right]^{\frac{1}{2}}$$

rather than ϕ_k. The appropriate Green function definition, instead of (4.19) would then be

$$D'(\mathbf{k}, t) = i\langle \Psi_0 | T(\phi'_{-k}(0)\phi'_k(t)) | \Psi_0 \rangle$$
$$= \frac{\omega(\mathbf{k})}{\omega_0(\mathbf{k})} D(\mathbf{k}, t) = -\frac{\omega(\mathbf{k})}{\omega^2 - (\omega(k) - i\eta)^2},$$

which is the standard form for a boson with frequency $\omega(\mathbf{k})$. The fact that we have chosen to use $D(\mathbf{k}, t)$ rather than $D'(\mathbf{k}, t)$ as the fundamental definition makes no difference, so long as we are consistent.

In a dense electron gas, we approximate $i\Lambda$ in $\kappa(\mathbf{k}, 0)$ by

$i\Lambda \simeq$ 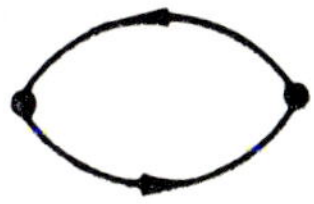 (4.35)

Then, for long wavelengths,

$$\omega(\mathbf{k}) \simeq \frac{\omega_{pl}}{(0{\cdot}66 r_s p_F{}^2)^{\frac{1}{2}}}|\mathbf{k}| \ . \tag{4.36}$$

The speed of sound, defined by

$$\omega(\mathbf{k}) = c|\mathbf{k}|$$

is given in this approximation by

$$c = \frac{\omega_{pl}}{(0.66 r_s p_F{}^2)^{\frac{1}{2}}} \simeq \frac{\omega_{pl}}{\nu_{pl}} \ \frac{v_F}{\sqrt{3}} \ , \tag{4.37}$$

where v_F is the electron velocity at the Fermi surface, and ω_{pl} and ν_{pl} are the ion and electron plasma frequencies.

Thus, the inclusion of self-energy corrections to the phonon propagator gives a phonon frequency proportional to $|\mathbf{k}|$ (for small $|\mathbf{k}|$), and, in fact, the estimate for the sound velocity is in reasonable agreement with experiment. With more care it is clear that we can get the scattering of phonons (and attenuation of sound waves) from the imaginary part of the pole of $D(\mathbf{k}, \omega)$, in analogy with the decay of plasmons, but we shall not go into this.

For $k = 2p_F$ W. Kohn,[3] recalling the infinite slope of κ as a function of $|\mathbf{k}|$, has predicted a corresponding behavior for $\omega(\mathbf{k})$ (and similarly for the energy of spin waves as a function of $\mathbf{k}$ in ferromagnetic metals[4]). These "Kohn effects" have now been observed.[5]

Now consider the screening of the electron-phonon interaction. In (4.30), if we again make approximation (4.35), then, for the high-density electron gas in the point-ion approximation, at long wavelengths,

$$\Gamma(k) = \frac{(4\pi e^2)^{\frac{3}{2}}(N/M)^{\frac{1}{2}}}{0{\cdot}66 r_s p_F{}^2 + k^2}k \tag{4.38}$$

replacing $g_k \sim k^{-1}$. Thus the inclusion of screening in this approximation already greatly reduces the electron-phonon interaction at long wavelengths where previously it was very large. This large reduction in the effective electron-phonon coupling strength is part of the justification for having omitted diagrams like 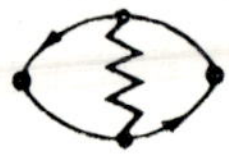 in the calculation of κ.

G. Effects of Phonons on Electrons and Holes

The presence of the phonon field has several effects on the electrons and holes. First there are the obvious ones of scattering the particles by absorption, emission or scattering of phonons. Second, the phonons can also modify the electron self-energy, the lowest-order effects arising from the contribution of

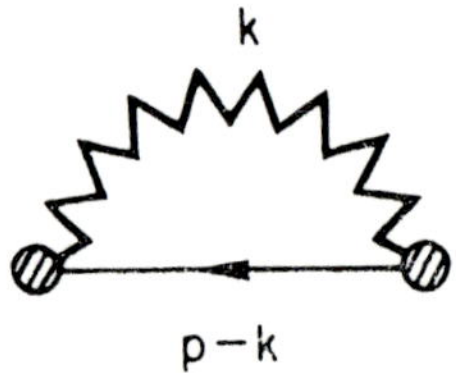

to $\Sigma(p)$. In computing this, we should of course use $D(k)$ for

and use Γ for each vertex (or the approximation to D and Γ that we have already introduced). This modification of electron self-energy first of all changes the real part of the electron energy and therefore the density of states at the Fermi surface, the specific heat, etc.[6] This correction is of the same order of magnitude as that due to electron-hole pair and plasmon effects, and must also be included if a realistic comparison with experiment is to be made. The self-energy also adds an imaginary part to the electron energy, because an additional mechanism for the decay of an electron or hole in the state **p** (phonon emission) has been introduced. Detailed calculation shows that this imaginary part is proportional to $p - p_F$ near the Fermi surface, and therefore dominates electron-electron scattering as a mode of decay. Unfortunately, the Landau theory of Fermi liquids, which seems so appealing, has required until now a $(p - p_F)^2$ dependence of the imaginary part of the self-energy for its justification, so that the development and justification of such a theory for an electron-phonon system is still an open problem.

A third effect of the phonons, as we have seen, is to introduce an additional effective interaction between electrons and/or holes. Thus in addition to diagram elements like

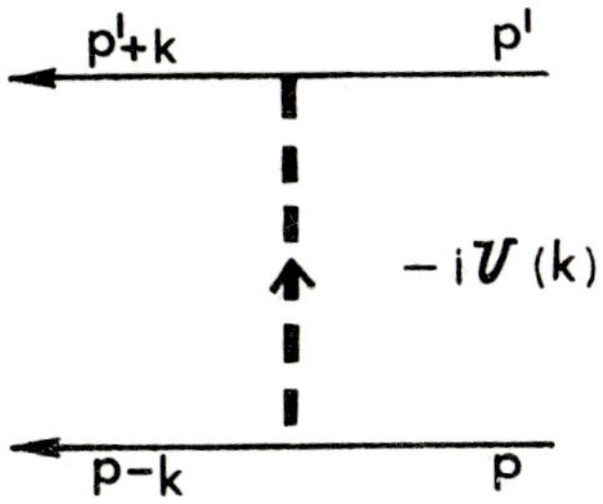

we now have elements like

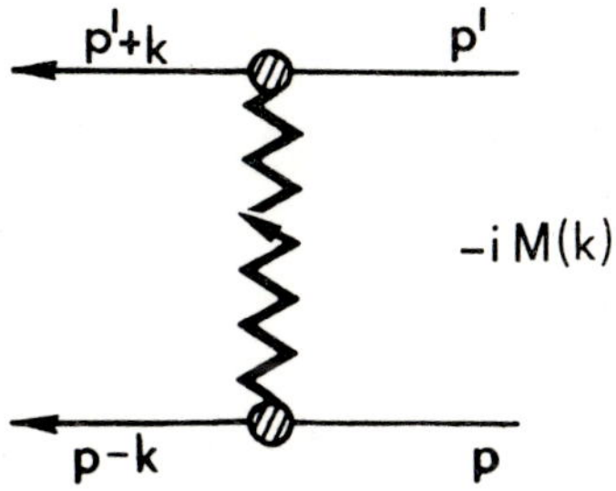

which implies an additional effective interaction between particles:

$$-iM(k) = -i[\Gamma(k)]^2 D(k) \simeq \frac{ig_k^2}{[\kappa(\mathbf{k},\,\omega)]^2}\left(\frac{\omega(\mathbf{k})}{\omega^2-\omega(\mathbf{k})^2}\right)\sqrt{\kappa(\mathbf{k},\,\omega)}. \tag{4.39}$$

If this element is the entire diagram, then energy conservation requires that $\omega = \epsilon(\mathbf{p}-\mathbf{k})-\epsilon(\mathbf{p})$. It is essentially this special interaction which is derived by Bardeen and Pines[7] and which serves as the basis of the superconductivity theory of Bardeen, Cooper and Schrieffer.[8] We see that in general, when the exchange of a phonon is only an element in a larger diagram, the effective interaction which this produces is more complicated, *all* values of ω playing a role.

Finally, a word of warning: One of the advantages of the diagrammatic analysis and perturbation theory is, as Du Bois has stressed, that the awkward subsidiary conditions in the Bohm-Pines method employed by Bardeen and Pines are replaced by the requirement that no graphs be double-counted, i.e., that one must avoid two partial sums simultaneously which involve the same kind of elements. I have been careful to avoid this in the partial sums already discussed. In going to higher order diagrams, one must *also* be careful to avoid double counting. For example, one must *not* include the diagram

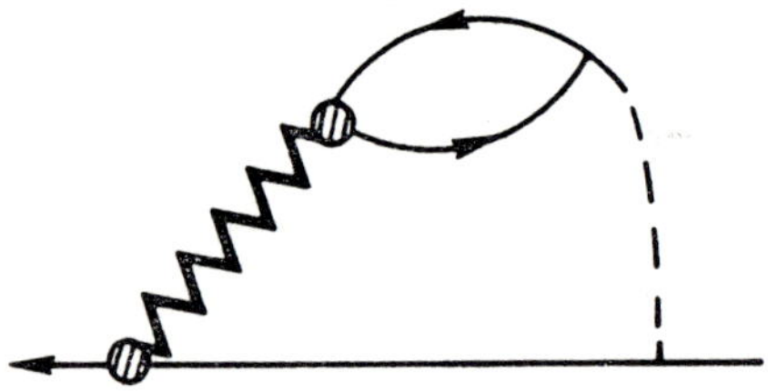

because this is already included in

That is all I shall say about the electron-phonon problem. It is clear that an extensive quantitative theory of phonon effects could be developed in much the same way that Du Bois has treated plasmon and electron-hole effects. In addition to the references given in Bardeen's *Handbuch* article and other references mentioned above, I might also suggest looking into Bogolubov, Tolmachev, and Shirkov[9] for a treatment of the many electron-phonon interactions, and also into Wentzel's[10] paper where Fourier components of the electron-hole density fluctuations and the phonons are treated as a coupled boson system.

REFERENCES

1. The dilute ion gas in a uniform background of electronic charge was first studied by K. Fuchs, *Proc. Roy. Soc.* (*London*) **A151**, 585 (1935); **A153**, 622 (1935); **A157**, 444 (1936). More recent work includes that of C. B. Clark, *Phys. Rev.* **109**, 1133 (1958); and A. A. Maradudin and G. H. Weiss, *Phys. Rev.* **123**, 1968 (1961). The dilute electron gas in a uniform background of ionic charge, which has many mathematical

similarities to the dilute ion gas, was first studied by E. P. Wigner, *Phys. Rev.* **46**, 1002 (1934) and *Trans. Faraday Soc.* **34**, 678 (1938). More recent work includes W. J. Carr, Jr., *Phys. Rev.* **122**, 1437 (1961); Rosemary Coldwell-Horsfall and A. Maradudin, *J. Math. Phys.* **1**, 385 (1960); and section II of Maradudin and Weiss, *op. cit.*

2. A. B. Migdal, *Soviet Phys. JETP* **7**, 996 (1958).
3. W. Kohn, *Phys. Rev. Letters* **2**, 393 (1959). See E. J. Woll, Jr. and W. Kohn, *Phys. Rev.* **126**, 1693 (1962) for detailed calculations.
4. E. J. Woll, Jr. and Stephen J. Nettel, *Phys. Rev.* **123**, 796 (1961).
5. B. N. Brockhouse, K. R. Rao, and A. D. B. Woods, *Phys. Rev. Letters* **7**, 93 (1961).
6. M. J. Buckingham and M. R. Schafroth, *Proc. Phys. Soc.* (*London*) **A67**, 828 (1954) and H. Ichimura, *Progr. Theoret. Phys.* **15**, 151 (1956).
7. J. Bardeen and D. Pines, *Phys. Rev.* **99**, 1140 (1955).
8. J. Bardeen, L. Cooper and J. R. Schrieffer, *Phys. Rev.* **108**, 1175 (1957).
9. N. N. Bogolubov, D. V. Shirkov, and V. V. Tolmachev, *A New Method in the Theory of Superconductivity*, New York Consultants Bureau, English Translation from Russian, 1959; and also slightly condensed in *Fort. d. Phys.* **6**, 605 (1958).
10. G. Wentzel, *Phys. Rev.* **108**, 1593 (1957).

CHAPTER V

MANY-FERMION SYSTEMS AT FINITE TEMPERATURES

I should now like to consider the generalization of the formalism already presented to finite temperatures. There are several ways to treat the finite temperature problem, too, but I shall restrict consideration to perturbation methods closely related to those already discussed. The initial published work from this point of view is that of Matsubara.[1] A direct development of Matsubara's approach is due to Abrikosov, Gor'kov, and Dzyaloshinskii, and it is this paper I shall follow most closely.[2]

A. Definition of G at Finite Temperatures

Recall the definition of G for zero temperature:

$$G(x, x') = -i\langle\Psi_0|T(\psi(x)\bar{\psi}(x'))|\Psi_0\rangle.$$

The generalization to finite temperatures is obtained by replacing the expectation in the ground state by an average over some statistical ensemble. A development for each of the statistical ensembles commonly used can be given,[3] but we shall work only with the grand canonical ensemble which is particularly convenient. The average of an operator $\mathcal{O}$ in the grand canonical ensemble is*

$$\langle\mathcal{O}\rangle = \frac{tr\ e^{-(\beta H-\alpha N)}\mathcal{O}}{tr\ e^{-(\beta H-\alpha N)}} = tr\ e^{\beta(\Omega-H+\mu N)}\mathcal{O}, \tag{5.1a}$$

where

$$e^{-\beta\Omega} \equiv tr\ e^{-(\beta H-\alpha N)} \tag{5.1b}$$

* Henceforth we assume the system occupies unit volume, which releases the symbol Ω for another conventional usage.

is the partition sum, and the trace goes over a complete set of states in which the number of particles is unrestricted. Ω, so defined, is a kind of free energy, the grand potential. The parameters $\beta = kT$ and $\alpha = \beta\mu$ characterize the ensemble. μ is the chemical potential, adjusted usually to give a desired value to the average number of particles.

Thus the Green function is

$$G^{\alpha,\beta}(x, x') = -i\, tr\, [e^{\beta(\Omega - H + \mu N)} T(\psi(x)\bar{\psi}(x'))] \ . \tag{5.2}$$

For systems that are time independent and spatially homogeneous one defines $G^{\alpha,\beta}(\mathbf{p}, t)$ and $G^{\alpha,\beta}(p) = G^{\alpha,\beta}(\mathbf{p}, \omega)$ in exact analogy with the $T = 0$ case. Henceforth, I shall supress superscripts α and β.

Just as for $T = 0$, we have three general concerns: to express desired physical quantities in terms of G (and perhaps two-particle G's), to examine the general properties of the Green functions, and to find methods for determining the Green functions. Let us consider the first of these questions.

B. Expressing Physical Quantities in Terms of the Green Functions

Several quantities can be expressed in terms of Green functions at finite temperatures in exactly the same way as at $T = 0$.

Average total number of particles:

$$\bar{N} = \left\langle \int d^3r\, \bar{\psi}(x)\psi(x) \right\rangle$$

$$= -i \int d^3r\, G(\mathbf{r}t, \mathbf{r}t^+) \ . \tag{5.3a}$$

For homogeneous systems, the average total number of particles can also be expressed in momentum space:

$$\bar{N} = -i \int \frac{d^3p}{(2\pi)^3} G(\mathbf{p}, t = 0^-) \ , \tag{5.3b}$$

suggesting that the average number of particles with momentum $\mathbf{p}$ is

$$\bar{n}_p = -iG(\mathbf{p}, t = 0^-)$$

The *average internal energy* can be similarly expressed. We found earlier that

$$H_1 = \frac{1}{2}\sum_p \left(iC_p{}^\dagger \frac{\partial C_p}{\partial t} - \epsilon_0(\mathbf{p})C_p{}^\dagger C_p\right) \tag{5.4a}$$

and so

$$H = \frac{1}{2}\sum_p \left(iC_p{}^\dagger \frac{\partial C_p}{\partial t} + \epsilon_0(\mathbf{p})C_p{}^\dagger C_p\right). \tag{5.4b}$$

Thus

$$\bar{E} \equiv \langle H \rangle = \frac{1}{2}\int \frac{d^3p}{(2\pi)^3}\left[\left(i\frac{\partial}{\partial t} + \epsilon_0(\mathbf{p})\right)G(\mathbf{p}, t)\right]_{t=0^-} \tag{5.5}$$

or, in coordinate space

$$\bar{E} = \frac{1}{2}\int d^3r \left[\left(i\frac{\partial}{\partial t} - \frac{\nabla_r{}^2}{2m}\right)G(\mathbf{r}t, \mathbf{r}'t')\right]_{r' \to r, t' \to t^+}$$

The grand potential can also be obtained from knowledge of the one-particle Green function in two different ways. The thermodynamic way is to derive the grand potential from a knowledge of $\bar{E}(\alpha, \beta)$ and $\bar{N}(\alpha, \beta)$. Both $\bar{N}$ and $\bar{E}$ can be expressed in terms of Ω:

$$\bar{N}(\alpha, \beta) = \frac{\partial}{\partial\alpha}\ln tr\, e^{-(\beta H - \alpha N)} = -\frac{\partial}{\partial\alpha}\beta\Omega(\alpha, \beta), \tag{5.6a}$$

$$\bar{E}(\alpha, \beta) = -\frac{\partial}{\partial\beta}\ln tr\, e^{-(\beta H - \alpha N)} = \frac{\partial}{\partial\beta}\beta\Omega(\alpha, \beta). \tag{5.6b}$$

Integration of these equations will give $\Omega(\alpha, \beta)$, although this may be very awkward, particularly if we only know $\bar{N}(\alpha, \beta)$ and $\bar{E}(\alpha, \beta)$ over a limited range of α and β.

An alternative procedure is to integrate on the coupling constant from zero to the actual value, as we did at $T = 0$. It turns out that a finite temperature analogue of the Feynman-Pauli theorem is valid. From (5.1b)

$$\Omega = -\frac{1}{\beta}\ln tr\, e^{-\beta H(g) + \alpha N},$$

where g is the interaction strength. Then,

$$\frac{\partial\Omega}{\partial g} = -\frac{1}{\beta}\frac{tr\frac{\partial}{\partial g}\,e^{-\beta H(g)+\alpha N}}{tr\;e^{-\beta H(g)+\alpha N}}\,.$$

To differentiate the exponential is not completely trivial, because the derivative of the exponent will not in general commute with the exponent itself. What saves us is the fact that we are only interested in a trace. Consider the general problem of finding

$$\frac{\partial}{\partial\lambda}tr\;\exp(K_0+\lambda K_1) = tr\frac{\partial}{\partial\lambda}\exp(K_0+\lambda K_1)\,,$$

where K_0 and K_1, both independent of λ, do not commute. Let us expand the exponential and differentiate term by term. A general term encountered is

$$tr\frac{1}{n!}(K_0+\lambda K_1)^{r-1}\frac{\partial(K_0+\lambda K_1)}{\partial\lambda}(K_0+\lambda K_1)^{n-r}$$
$$= tr\frac{1}{n!}(K_0+\lambda K_1)^{n-1}\frac{\partial(K_0+\lambda K_1)}{\partial\lambda},$$

because the trace of a product of operators is not changed if we make a cyclic permutation of the operators. Thus in every term, we can bring the differentiated term to the right, obtaining after summing,

$$\frac{\partial}{\partial\lambda}tr\;e^{K_0+\lambda K_1} = tr\;e^{(K_0+\lambda K_1)}K_1\,.$$

If now we assume $H = H_0+H_1$ with H_1 linear in a coupling constant g, then

$$\frac{\partial H}{\partial g} = \frac{H_1}{g}$$

and

$$\frac{\partial\Omega}{\partial g} = \frac{1}{g}\langle H_1\rangle\,. \tag{5.7}$$

Integrating from $g = 0$ to the actual value, we obtain

$$\Omega(g) = \Omega(0)+\int_0^g\frac{dg'}{g'}\langle H_1\rangle_{g'}\,. \tag{5.8}$$

$\langle H_1 \rangle_{g'}$ can be expressed in terms of the one-particle Green function from (5.4a) in analogy with (5.5).

The responses to electric and magnetic fields can also be expressed in terms of two-particle Green functions, or equivalently in terms of the polarization propagator and other similar correlation functions defined for finite temperatures. A general treatment is given by Martin and Schwinger, Section IV.

C. Free One-Particle Green Functions

It is easy to compute $G_0(\mathbf{p}, t)$, a Green function for a noninteracting system. By definition

$$G_0(\mathbf{p}, t) = -i\langle T(C_p(t)C_p^\dagger(0))\rangle \, , \tag{5.9}$$

where the ensemble is now of systems of free particles and

$$C_p(t) = e^{-i\epsilon_0(\mathbf{p})t}C_p(0) \; .$$

For $t > 0$,

$$G_0(\mathbf{p}, t) = -i\, e^{-i\epsilon_0(\mathbf{p})t}(1-\bar{n}_p) \tag{5.10a}$$

and for $t < 0$,

$$G_0(\mathbf{p}, t) = i\, e^{-i\epsilon_0(\mathbf{p})t}\bar{n}_p, \tag{5.10b}$$

where

$$\bar{n}_p = \frac{tr\; e^{-\beta(H-\mu N)}C_p{}^\dagger C_p}{tr\; e^{-\beta(H-\mu N)}} = \frac{1}{e^{\beta\epsilon_0(\mathbf{p})-\alpha}+1} \, . \tag{5.11}$$

We can also compute $G_0(p)$ defined by

$$G_0(p) = \int_{-\infty}^{\infty} dt\; \exp(iEt) G_0(\mathbf{p}, t)$$

$$= \frac{1-{}_p}{E-\epsilon_0(\mathbf{p})+i\eta} + \frac{\bar{n}_p}{E-\epsilon_0(\mathbf{p})-i\eta} \tag{5.12}$$

Separating real and imaginary parts,

$$G_0(p) = \mathscr{P}\frac{1}{E-\epsilon_0(\mathbf{p})} - \pi i\delta\{E-\epsilon_0(\mathbf{p})\}(1-2\bar{n}_p)$$

$$= \mathscr{P}\frac{1}{E-\epsilon_0(\mathbf{p})} - \pi i\delta\{E-\epsilon_0(\mathbf{p})\} \tanh\left[\frac{\beta\epsilon_0(\mathbf{p})-\alpha}{2}\right] \tag{5.13}$$

From (5.12) it is also evident that $G_0(p)$ has poles in both the upper and lower half-planes. Its analytic properties are therefore more complicated than for $T = 0$.

D. Exact One-Particle Green Functions

1. *Spectral Representation*

Let us assume a homogeneous time-independent system and consider the exact one-particle Green function $G(\mathbf{p}, E)$ defined by

$$G(\mathbf{p}, E) = \int_{-\infty}^{\infty} dt \exp(iEt)G(\mathbf{p}, t) . \tag{5.14}$$

Introducing the definition of $G(\mathbf{p}, t)$ and expanding in a complete set of states, we obtain

$$\begin{aligned} G(\mathbf{p}, E) = -i \int_0^{\infty} dt \exp(iEt) \sum_{nm} \exp(\beta\Omega) \\ \langle n|\exp(-\beta H+\alpha N)C_p(t)|m\rangle\langle m|C_p{}^{\dagger}|n\rangle \\ +i\int_{-\infty}^{0} dt \exp(iEt) \sum_{nm} \exp(\beta\Omega) \\ \langle m|\exp(-\beta H+\alpha N)\; C_p{}^{\dagger}|n\rangle\langle n|C_p(t)|m\rangle. \end{aligned}$$

Observe that if $\langle m|C_p{}^{\dagger}|n\rangle$ is to be different from zero, the state $|n\rangle$ must have one less particle than the state $|m\rangle$, so that

$$e^{-\beta\mathscr{E}_m+\alpha N_m} = e^{-\beta\mathscr{E}_m+\alpha(N_n+1)} .$$

Thus,

$$\begin{aligned} G(\mathbf{p}, E) = \sum_{mn} e^{\beta\Omega}\Big\{-i\int_0^{\infty} dt\, e^{iEt}\, e^{i(\mathscr{E}_n-\mathscr{E}_m+i\eta)t}\, e^{-\beta\mathscr{E}_n+\alpha N_n} , \\ +i\int_{-\infty}^{0} dt\, e^{iEt}\, e^{i(\mathscr{E}_n-\mathscr{E}_m-i\eta)t}\, e^{-\beta\mathscr{E}_m+\alpha(N_n+1)}\Big\}|\langle n|C_p|m\rangle|^2 , \end{aligned}$$

where we have introduced η's to insure convergence. After integrating on t,

$$\begin{aligned} G(\mathbf{p}, E) = \sum_{mn} \exp(\beta\Omega)\left\{\frac{\exp(-\beta\mathscr{E}_n+\alpha N_n)}{E+\mathscr{E}_n-\mathscr{E}_m+i\eta} + \frac{\exp\{-\beta\mathscr{E}_m+\alpha(N_n+1)\}}{E+\mathscr{E}_n-\mathscr{E}_m-i\eta}\right\} \\ \times|\langle n|C_p|m\rangle|^2. \end{aligned} \tag{5.15}$$

Introduce the spectral function $\rho(\mathbf{p}, E')$:

$$\rho(\mathbf{p}, E')dE' = \sum_{nm} e^{\beta\Omega}[e^{-\beta\mathscr{E}_n+\alpha N_n}+e^{-\beta\mathscr{E}_m+\alpha N_m}]\,|\langle n|C_p|m\rangle|^2,$$

$$E' < \mathscr{E}_m-\mathscr{E}_n < E'+dE'. \qquad (5.16)$$

Then

$$G(\mathbf{p}, E) = \mathscr{P}\int_{-\infty}^{\infty}\frac{\rho(\mathbf{p}, E')dE'}{E-E'}-\pi i\, e^{\beta\Omega}\sum_{mn}\delta(E-\mathscr{E}_m+\mathscr{E}_n)$$
$$\times[e^{-\beta\mathscr{E}_n+\alpha N_n} - e^{-\beta\mathscr{E}_m+\alpha(N_n+1)}]\,|\langle n|C_p|m\rangle|^2 .$$

But if $\mathscr{E}_m-\mathscr{E}_n = E$, then

$$e^{-\beta\mathscr{E}_n+\alpha N_n} - e^{-\beta\mathscr{E}_m+\alpha(N_n+1)} = \tanh\frac{\beta E-\alpha}{2}$$
$$\times[e^{-\beta\mathscr{E}_n+\alpha N_n} + e^{-\beta\mathscr{E}_m+\alpha(N_n+1)}], \qquad (5.17)$$

so that

$$G(\mathbf{p}, E) = \mathscr{P}\int_{-\infty}^{\infty}\frac{\rho(\mathbf{p}, E')dE'}{E-E'}$$
$$-\pi i\tanh\frac{\beta E-\alpha}{2}\int_{-\infty}^{\infty}dE'\rho(\mathbf{p}, E')\delta(E-E'), \qquad (5.18)$$

the spectral representation of $G(\mathbf{p}, E)$.

2. *Dispersion Relation*

From this spectral representation, taking imaginary parts, we have

$$Im\, G(\mathbf{p}, E) = -\pi\tanh\frac{\beta E-\alpha}{2}\rho(\mathbf{p}, E),\ \text{for } E \text{ real.} \qquad (5.19a)$$

Taking real parts and using (5.19a) we obtain a dispersion relation

$$Re\, G(\mathbf{p}, E) = \mathscr{P}\int_{-\infty}^{\infty}\frac{\rho(\mathbf{p}, E')dE'}{E-E'}$$
$$= -\frac{1}{\pi}\mathscr{P}\int_{-\infty}^{\infty}\coth\frac{\beta E'-\alpha}{2}\,\frac{Im\, G(\mathbf{p}, E')}{E-E'}dE' . \qquad (5.19b)$$

3. *Properties of the Spectral Function* $\rho(\mathbf{p}, E)$

Note two properties possessed by $\rho(\mathbf{p}, E)$ and which should therefore be possessed by any approximation to $\rho(\mathbf{p}, E)$:

$$\rho(\mathbf{p}, E) \geqslant 0 \tag{5.20a}$$

and

$$\begin{aligned}\int dE\,\rho(\mathbf{p}, E) &= \exp(\beta\Omega)\sum_{nm}\{\exp(-\beta\mathscr{E}_n+\alpha N_n)+\exp(-\beta\mathscr{E}_m+\alpha N_m)\}\,|\langle n|C_p|m\rangle|^2 \\ &= e^{\beta\Omega}\,tr[e^{-\beta H+\alpha N}(C_pC_p{}^\dagger+C_p{}^\dagger C_p)] \\ &= 1.\end{aligned} \tag{5.20b}$$

These properties of $\rho(\mathbf{p}, E)$, together with the spectral representation (5.18) or the equivalent relations (5.19) serve to characterize the analytic properties of $G(p)$. This may be very useful if we are able to construct $G(p)$ for imaginary E, let us say, and we wish to continue it analytically to real E.

E. Shifting the Single-Particle Energy so $\mu = 0$

In a later section we wish to develop the perturbation theory for constructing the various Green functions. It turns out to be more convenient for this to start by considering a slightly different Green function which we now define.

Suppose we measure the one-particle energies from the energy μ, now considered to be fixed. The old Hamiltonian

$$H = \int d^3r\,\bar\psi(\mathbf{r})\left(-\frac{\nabla^2}{2m}\right)\psi(\mathbf{r}) + \int v\,\bar\psi\bar\psi\psi\psi$$

is now replaced by a new Hamiltonian, which depends on μ:

$$\begin{aligned}K &= \int d^3r\,\bar\psi(\mathbf{r})\left(-\frac{\nabla^2}{2m}-\mu\right)\psi(\mathbf{r})+\int v\,\bar\psi\bar\psi\psi\psi \\ &= H-\mu N\end{aligned} \tag{5.21}$$

Now suppose for this Hamiltonian we define $\bar G$ by

$$\bar G(x, x') = \frac{-i\,tr\exp(-\beta K)T(\psi(x)\bar\psi(x'))}{tr\exp(-\beta K)}, \tag{5.22}$$

where now

$$\psi(x) = \exp(iKt)\psi(\mathbf{r}0)\exp(-iKt), \text{ etc.} \tag{5.23}$$

and the trace is still taken over states with all numbers of particles. This Green function differs from G only in that the Heisenberg operators ψ and $\bar{\psi}$ now develop in time according to K rather than H. Because $[H, N] = 0$, it is trivial to show that

$$\bar{G}(x, x') = e^{-i\mu(t'-t)}G(x, x'). \tag{5.24a}$$

and therefore, for homogeneous time-independent systems,

$$\bar{G}(\mathbf{p}, E-\mu) = G(\mathbf{p}, E) . \tag{5.24b}$$

We shall now work with $\bar{G}$ rather than G.

F. Boundary Conditions for Green Functions—Imaginary Time Green Functions

The finite temperature Green function $G(x, x')$ and the one at absolute zero are both averages of the operator $T(\psi(x)\bar{\psi}(x'))$ as we've seen, the averages simply being over different ensembles. It is possible in either case to derive an infinite hierarchy of coupled equations connecting the Green functions of one, two, etc., particles. These are just the appropriate averages of a set of coupled operator equations involving the operators $T(\psi(x)\bar{\psi}(x'))$, $T(\psi(x_1)\psi(x_2)\bar{\psi}(x_2')\bar{\psi}(x_1'))$, etc. I won't write these down but refer you to Martin and Schwinger. The important thing is that the equations for the Green functions at any temperature including $T = 0$ are exactly the same, since they are all just various averages of the same operator equations. What distinguishes the solutions appropriate to one temperature from those appropriate to another is the boundary conditions we require to be satisfied. One way to satisfy the boundary conditions is to convert the equations to integral equations that incorporate the required conditions. This can be done by using the appropriate noninteracting Green functions for the desired temperature. Another way is to expand the Green functions in a complete set of functions, all satisfying the required boundary conditions.

Because the boundary conditions are so crucial, let us see what they are. Just for orientation, assume $t > t'$ and let us use the cyclic property of a trace (that it is unchanged by a cyclic permutation of the operators). Then

$$\begin{aligned}\bar{G}(\mathbf{r}t,\mathbf{r}'t') &= -\frac{i\,tr\exp(-\beta K)\psi(x)\bar{\psi}(x')}{tr\exp(-\beta K)}\\ &= -\frac{i\,tr\,e^{-(\beta-it)K}\psi(\mathbf{r})\,e^{-i(t-t')K}\bar{\psi}(\mathbf{r}')\,e^{-it'K}}{tr\exp(-\beta K)}\\ &= -\frac{i\,tr\,e^{-\beta K}[e^{(\beta+it')K}\bar{\psi}(\mathbf{r}')\,e^{-(\beta+it')K}][e^{itK}\psi(\mathbf{r})\,e^{-itK}]}{tr\exp(-\beta K)}\\ &= -G(\mathbf{r}t,\mathbf{r}'t'-i\beta)\,,\end{aligned} \tag{5.25}$$

if in some sense we could say that $t-i\beta > t$. This would be a useful sort of boundary condition.

This suggests one of two possible procedures:

(1) We could let $\beta \to i\gamma(1-i\eta)$, γ real, consider Green functions for imaginary temperature, and having calculated them, continue analytically to real temperature. This is essentially the procedure followed by Martin and Schwinger.

(2) Alternatively, we could let $it = \tau$ and $it' = \tau'$ understanding the T product to be ordered according to the relative sizes of τ and τ'. Then eventually we would have to continue analytically to real times (or real time differences). This is the procedure of Abrikosov *et al.* and the one we shall follow.

To be systematic, we introduce still another Green function defined most conveniently directly in terms of imaginary times:

$$\mathcal{G}(\mathbf{r}\tau,\mathbf{r}'\tau') = -\frac{tr\exp(-\beta K)T\{\psi(\mathbf{r}\tau)\bar{\psi}(\mathbf{r}'\tau')\}}{tr\exp(-\beta K)}, \tag{5.26}$$

where now

$$\psi(\mathbf{r}\tau) = \exp(K\tau)\psi(\mathbf{r})\exp(-K\tau),\ \text{etc.} \tag{5.27}$$

Consider homogeneous time-independent systems (this is not necessary but simplifies the discussion—see Martin and Schwinger for a more general treatment). Then just as for (5.25), it is easy to show

$$\mathcal{G}(\mathbf{r}-\mathbf{r}',\tau-\tau') = -\mathcal{G}(\mathbf{r}-\mathbf{r}',\tau-\tau'-\beta)$$

if $\tau-\tau' > 0$ and $\tau-\tau'-\beta < 0$, and

$$\mathcal{G}(\mathbf{r}-\mathbf{r}',\tau-\tau') = -\mathcal{G}(\mathbf{r}-\mathbf{r}',\tau-\tau'+\beta)$$

if $\tau-\tau' < 0$ and $\tau-\tau'+\beta > 0$. Letting $\mathbf{r}-\mathbf{r}' \to \mathbf{r}$ and $\tau-\tau' \to \tau$ these are rewritten

$$\mathscr{G}(\mathbf{r}, \tau) = -\mathscr{G}(\mathbf{r}, \tau-\beta) \tag{5.28a}$$

for $\tau > 0$ and $\tau-\beta < 0$, and

$$\mathscr{G}(\mathbf{r}, \tau) = -\mathscr{G}(\mathbf{r}, \tau+\beta) \tag{5.28b}$$

for $\tau < 0$ and $\tau+\beta > 0$.

Given $\mathscr{G}$ for $-\beta < \tau < 0$, we thus define $\mathscr{G}$ for $0 < \tau < \beta$. It is convenient to define $\mathscr{G}$ outside the interval $(-\beta, \beta)$ by requiring it to be periodic with period 2β.

Expanding $\mathscr{G}(\mathbf{r}, \tau)$ in a Fourier series

$$\mathscr{G}(\mathbf{r}, \tau) = \frac{1}{\beta}\sum_{n=-\infty}^{\infty} \exp(-iE_n\tau)\mathscr{G}(\mathbf{r}, E_n), \tag{5.29}$$

we must require

$$E_n = \frac{(2n+1)\pi}{\beta}, \tag{5.30}$$

so that $\mathscr{G}(\mathbf{r}, \tau)$ is periodic in τ with period 2β and so that it satisfies (5.28).*

The expansion coefficients $\mathscr{G}(\mathbf{r}, E_n)$ are the analogue of the function $G(\mathbf{r}, E)$ at absolute zero. Inverting (5.29), we have

$$\mathscr{G}(\mathbf{r}, E_n) = \frac{1}{2}\int_{-\beta}^{\beta} \mathscr{G}(\mathbf{r}, \tau)\exp(iE_n\tau)d\tau\ . \tag{5.31}$$

The momentum Green functions for noninteracting particles are particularly useful in the perturbation theory. They are readily calculated as

$$\begin{aligned}\mathscr{G}_0(\mathbf{p}, \tau) &= -\frac{e^{-(\epsilon_0(\mathbf{p})-\mu)\tau}}{1+e^{-\beta(\epsilon_0(\mathbf{p})-\mu)}}, \quad \beta > \tau > 0\\ &= -\frac{e^{-(\epsilon_0(\mathbf{p})-\mu)\tau}}{1+e^{\beta(\epsilon_0(\mathbf{p})-\mu)}}, \quad -\beta < \tau < 0\end{aligned} \tag{5.32}$$

and

$$\mathscr{G}_0(\mathbf{p}, E_n) = \frac{1}{iE_n-\{\epsilon_0(\mathbf{p})-\mu\}}\ . \tag{5.33}$$

* A similar development can be given for bosons (e.g. phonons). One finds $\mathscr{G}(\mathbf{r},\tau) = +\mathscr{G}(\mathbf{r}, \tau-\beta)$ for (5.28a) and so $E_n = (2n\pi/\beta)$ for (5.30).

G. Perturbation Theory and Diagrammatic Analysis

The perturbation theory can be developed for $\mathscr{G}$ just as it was developed for real time differences at $T = 0$. We must now be a little more careful because the operator $\exp(K\tau)$ which generates imaginary time displacements is not unitary. I shall sketch the development only briefly.

1. *The Interaction Picture*

First, we introduce the interaction picture assuming the decomposition

$$K = K_0 + K_1 .$$

Then define

$$\psi^I(\tau) = e^{K_0\tau}\psi(0)e^{-K_0\tau}$$

which together with

$$\psi^H(\tau) = e^{K\tau}\,\psi(0)\,e^{-K\tau}$$

implies

$$\psi^H(\tau) = e^{K\tau}\,e^{-K_0\tau}\,\psi^I(\tau)\,e^{K_0\tau}\,e^{-K\tau} . \tag{5.34}$$

One can show that

$$e^{-K\tau} = e^{-K_0\tau}\,T(e^{-\int_0^\tau K_1(s)ds}) \tag{5.35a}$$

and

$$e^{K\tau} = \widetilde{T}\,e^{\int_0^\tau K_1(s)ds}\,e^{K_0\tau}, \tag{5.35b}$$

where

$$K_1(s) = \exp(K_0 s)K_1 \exp(-K_0 s)$$

and T and $\widetilde{T}$ mean to order all operators according to their arguments increasing from right to left (for T) or from left to right (for $\widetilde{T}$).

Then we may define

$$\begin{aligned} U(\tau, \tau') &\equiv e^{K_0(\tau-\tau')}\,e^{-K(\tau-\tau')} \\ &= T(e^{-\int_{\tau'}^{\tau} K_1(s)ds}) \text{ for } \tau > \tau' \end{aligned} \tag{5.36a}$$

and

$$U(\tau, \tau') \equiv \{U(\tau', \tau)\}^{-1} = e^{-K_0(\tau-\tau')}$$
$$= \tilde{T}\left\{\exp\left(-\int_\tau^{\tau'} K_1(s)\,ds\right)\right\} \text{ for } \tau < \tau'. \quad (5.36b)$$

It is then easy to show that

$$U(\tau_1\tau_2)U(\tau_2\tau_3) = U(\tau_1\tau_3) \quad (5.37)$$

for any order of the "times" τ_1, τ_2, τ_3.

Expressing ψ and $\bar{\psi}$ in (5.26) terms of U, ψ^I and $\bar{\psi}^I$, and using (5.35a) for $\exp(-K\beta)$, it is then also easy to show that

$$\mathscr{G}(\mathbf{r}\tau, \mathbf{r}'\tau') = -\frac{tr\,\exp(-\beta K_0)T(\psi^I(\mathbf{r}\tau)\bar{\psi}(\mathbf{r}'\tau')S(\beta))}{tr\,\exp(-\beta K_0)S(\beta)} \quad (5.38)$$

where

$$S(\beta) = U(\beta, 0) = T\{e^{\int_0^\beta K_1(s)ds}\}, \quad (5.39)$$

and the T-product is understood to include a power series expansion of S followed by a "time" ordering of all factors following T. We see that the generalization of $\langle \Phi_0|\mathcal{O}|\Phi_0 \rangle$ to finite temperatures is $tr\,\exp(-\beta K_0)\mathcal{O}/tr\,\exp(-\beta K_0)$.

2. *Normal Products and Contractions*

Wick's theorem holds for the terms of the expansion implicit in (5.38), providing we define the normal product of a set of operators in a suitable way. We start by generalizing the decompositions of (3.44a, b):

$$\psi(\mathbf{r}\tau) = \underbrace{\sum_p \lambda_p \exp(i\mathbf{p}\cdot\mathbf{r})c_p(\tau)}_{\equiv\,\psi^+(\mathbf{r}\tau)} + \underbrace{\sum_p (1-\lambda_p)\exp(i\mathbf{p}\cdot\mathbf{r})c_p(\tau)}_{\equiv\,\psi^-(\mathbf{r}\tau)} \quad (5.40a)$$

$$\bar{\psi}(\mathbf{r}\tau) = \underbrace{\sum_p (1-\lambda_p)\exp(-i\mathbf{p}\cdot\mathbf{r})c_p{}^\dagger(\tau)}_{\equiv\,\bar{\psi}^-(\mathbf{r}\tau)} + \underbrace{\sum_p \lambda_p \exp(-i\mathbf{p}\cdot\mathbf{r})c_p{}^\dagger(\tau)}_{\equiv\,\bar{\psi}^+(\mathbf{r}\tau)}. \quad (5.40b)$$

The coefficients λ_p are still to be determined. In terms of this decomposition, we define the normal product of a set of field operators exactly as in Section III*F*1. To determine the λ_p's, we require that

$$\langle N(\bar{\psi}(\mathbf{r}\tau)\psi(\mathbf{r}'\tau'))\rangle = 0 \ . \tag{5.41}$$

Direct substitution of (5.40 a, b) in (5.41) yields an equation for λ_p:

$$\lambda_p{}^2 = \langle n_p\rangle = f_p \ . \tag{5.42}$$

Now f_p is the Fermi function at temperature T. It is obvious that the normal products $N(\bar{\psi}\bar{\psi})$ and $N(\psi\psi)$ also have zero ensemble averages. What is not obvious (in contrast to the $T = 0$ case) is the fact that the ensemble averages of normal products of four, six, etc. field operators vanish. In fact, Matsubara in his original paper believed it to be true only in the limit of infinite volume. I shall not prove this theorem here.

The contraction is again defined as

$$\underbrace{AB} = T(AB)-N(AB), \tag{5.43}$$

and is a c-number. Taking the ensemble average of this equation, we obtain a way to compute various contractions:

$$\underbrace{AB} = \langle T(AB)\rangle \ . \tag{5.44}$$

We list the important contractions:

$$\underbrace{\psi^I(\mathbf{r}\tau)\bar{\psi}^I(\mathbf{r}'\tau')} = -\mathscr{G}_0(\mathbf{r}\tau,\mathbf{r}'\tau') \ , \tag{5.45a}$$

$$\underbrace{\psi^I(\mathbf{r}\tau)\bar{\psi}^I(\mathbf{r}\tau)} = \rho(\mathbf{r}\tau) \ , \tag{5.45b}$$

$$\underbrace{\psi^I(\mathbf{r}\tau)\psi^I(\mathbf{r}'\tau')} = \underbrace{\bar{\psi}^I(\mathbf{r}\tau)\bar{\psi}^I(\mathbf{r}'\tau')} = 0 \ . \tag{5.45c}$$

Wick's decomposition of a T-product into N-products times contractions now implies that the ensemble average of a T-product is the sum over all pairings, with appropriate factors of (-1), of the product of contractions of the pairs.* These results are generalizable

*This result, or closely similar results, have been proved by Bloch and de Dominicis, *Nucl. Phys.* **7,** 459 (1958); D. J. Thouless, *Phys. Rev.* **107,** 1162 (1957); and M. Gaudin, *Nucl. Phys.* **15,** 89 (1960).

to bosons in the obvious way, although the proofs are again not obvious.

3. *Diagrams and Rules*

The diagrammatic analysis in $\mathbf{r}\tau$-space goes through just as in x-space at $T=0$, except that $-\mathscr{G}_0(\mathbf{r}-\mathbf{r}', \tau-\tau')$ is associated with a directed particle line (for homogeneous time-independent systems), and $\rho(\mathbf{r}\tau)$ is associated with a single line closed on itself. It is possible to make the τ-integrations more symmetric if it is observed that each vertex always has an even number of particle lines meeting there. Thus, instead of taking each τ-integral from 0 to β, we can take one half the integral from $-\beta$ to β. The range from $-\beta$ to 0 gives the same as the range from 0 to β, multiplied by an even power of (-1). This facilitates introduction of $\mathscr{G}(\mathbf{r}, E_n)$ by (5.31).

In going over to momentum diagrams, the particle lines carry $-\mathscr{G}(\mathbf{p}, E_n)$ and each vertex has a delta function of the sum of the 4-momenta entering the vertex minus those leaving it. For the 4th components of momentum, the delta function is a Kronecker delta rather than a Dirac delta function. We must integrate $\int d^3p/(2\pi)^3$ over all three momenta and sum $\beta^{-1}\sum_n$ over all fourth components.

The counting, partial summations, Dyson equations, etc., all follow exactly as at $T=0$.

In going to the limit $T\to 0$ or $\beta\to\infty$, the variable E_n becomes a continuous variable. Summations like $\beta^{-1}\sum_n f(E_n)$ are Riemann sums for the integrals $(2\pi)^{-1}\int_{-\infty}^{\infty} dE f(E)$. Let us introduce variables $E'=iE$ so that factors like $(iE-\epsilon_0(\mathbf{p})+\mu)^{-1}$ go into $(E'-\epsilon_0(\mathbf{p})+\mu)^{-1}$. The E-integrations now go into

$$\frac{-i}{2\pi}\int_{-i\infty}^{i\infty} dE'$$

integrations which can be deformed into

$$\frac{-i}{2\pi}\int_{\lrcorner\!\!\nearrow} dE'$$

integrations, so that, aside from differences in the origin of energy (taken to be μ for $\mathscr{G}$) and factors of i introduced in the definition of $\mathscr{G}$, it is obvious that $T \neq 0$ contributions go over into $T=0$ contributions.

There is one exception to this rule, the class of so-called anomalous diagrams first discussed by Kohn and Luttinger.[4] As we have remarked earlier, diagrams containing a subdiagram like

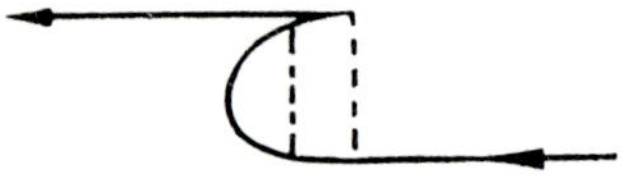

do not contribute in the $T = 0$ theory, but do contribute for $T \neq 0$. If the limit of infinite volume is taken *before* the limit $T \to 0$ (and Kohn and Luttinger show that it must be, to obtain sensible results), the contributions of anomalous diagrams survive. These contributions, affecting the one-particle self-energies, can be thought to contribute to the chemical potential μ. Until now we have been considering μ as a fixed parameter, but when perturbation theory to any order in the interaction is carried out, if $\langle N \rangle$ is to remain fixed then μ will depend on the strength of the interaction. In the limit $T \to 0$, one might well ask if μ, so calculated, tends to the value one would calculate in the $T = 0$ theory, neglecting anomalous diagrams. Also, does the internal energy tend to the ground state energy calculated in a $T = 0$ theory? Luttinger and Ward[5] have shown, to all orders in perturbation theory, that the $T = 0$ theory without anomalous diagrams is indeed the limit of the $T \neq 0$ theory with these diagrams providing (a) the unperturbed single-particle energy is a spherically symmetric function in **k**-space, and (b) the particles either have spin $\frac{1}{2}$ or have no spin-dependent interactions. Brout and Englert[6] have confirmed this theorem by developing a linked cluster expansion for the canonical ensemble, where no chemical potential appears in the formalism, but where the diagrams contain additional complications.

H. Relation Between $\mathcal{G}$ and $\bar{G}$

It is apparent from the definitions that once we know $\mathcal{G}(\mathbf{r}, \tau)$ for $-\beta < \tau < \beta$, we can get $\bar{G}(\mathbf{r}, \tau)$ by formally replacing τ by it, and multiplying by i. We can then let $|t|$ be arbitrarily large.

A more subtle problem is to construct $\bar{G}(\mathbf{p}, E)$ knowing $\mathcal{G}(\mathbf{p}, E_n)$. The key to this problem is to examine the analytic properties of $\mathcal{G}$

and $\bar{G}$ using the Lehmann representation. Paralleling the development leading to (5.18), it is easy to show that

$$\mathscr{G}(\mathbf{p}, E_n) = \int_{-\infty}^{\infty} dE' \frac{\rho(\mathbf{p}, E')}{iE_n + \mu - E'} \tag{5.46}$$

and

$$\begin{aligned} \bar{G}(\mathbf{p}, E) &= G(\mathbf{p}, E+\mu) \\ &= \mathscr{P} \int dE' \frac{\rho(\mathbf{p}, E')}{E+\mu-E'} \\ &\quad - \pi i \tanh \frac{\beta E}{2} \int dE' \rho(\mathbf{p}, E') \delta(E+\mu-E') , \end{aligned} \tag{5.47}$$

where $\rho(\mathbf{p}, E)$ is defined by (5.16).

First observe from (5.46) that $\mathscr{G}(\mathbf{p}, E_n)$ considered as a function of the continuous variable E_n is analytic in both the right and left half-planes of E_n but it has a discontinuity in its imaginary part as ReE_n passes through zero, i.e., the imaginary axis is a branch line for $\mathscr{G}(\mathbf{p}, E_n)$. To find $\bar{G}(\mathbf{p}, E)$, let us start then with $\mathscr{G}(\mathbf{p}, E_n)$ for $E_n > 0$ and let $E_n \to -iE+\eta$ (so as not to cross the cut). Then

$$\mathscr{G}(\mathbf{p}, -iE+\eta) = \mathscr{P} \int dE' \frac{\rho(\mathbf{p}, E')}{E+\mu-E'} - \pi i \int dE' \rho(\mathbf{p}, E') \delta(E+\mu-E'), \tag{5.48}$$

so that

$$Re\, \bar{G}(\mathbf{p}, E) = Re\, \mathscr{G}(\mathbf{p}, -iE+\eta) \tag{5.49a}$$

$$Im\, \bar{G}(\mathbf{p}, E) = \tanh \frac{\beta E}{2} Im\, \mathscr{G}(\mathbf{p}, -iE+\eta) \tag{5.49b}$$

It is a simple exercise to check that these relations hold true for $\mathscr{G}_0$ and $\bar{G}_0$.

References

1. T. Matsubara, *Progr. Theoret. Phys.* **14**, 351 (1955).
2. A. A. Abrikosov, L. P. Gor'kov, and I. E. Dzyaloshinskii, *Soviet Phys. JETP* **9**, 636 (1959).

   A much more extensive paper in the same direction is the work of P. C. Martin and J. Schwinger, *Phys. Rev.* **115**, 1342 (1959). This paper obtains many results, some extremely formal, and includes a great deal not discussed by Abrikosov, Gor'kov, and Dzyaloshinskii. It is particularly thorough on applications of Green's functions and nonperturbational approaches for determining the Green functions. Also very formal, and along similar lines, are the three papers of E. S. Fradkin, *Soviet Phys. Dokl.* **4**, 327 and 347 (1959), and *Soviet Phys. JETP* **9**, 912 (1959). The last paper, also available in *Nuclear Phys.* **12**, 465 (1959), is a more extended and generalized version of the briefer reports in Doklady. Also along these lines is V. L. Bonch-Bruevich and Sh. M. Kogan, *Ann. Phys.* **9**, 125, 1960. See also the three reviews and the book listed in the bibliography.

Somewhat different points of view are taken by C. Bloch and C. de Dominicis, *Nuclear Phys.* **7**, 459 (1958); **10**, 181 (1959); and **10**, 509 (1959); by E. Montroll and J. Ward in *Phys. of Fluids*, **1**, 55 (1958); and others.

3. Martin and Schwinger, *op. cit.* also consider the microcanonical ensemble.
4. W. Kohn and J. M. Luttinger, *Phys. Rev.* **118**, 41 (1960).
5. J. M. Luttinger and J. C. Ward, *Phys. Rev.* **118**, 1417 (1960).
6. R. Brout and F. Englert, *Phys. Rev.* **120**, 1519 (1960).

BIBLIOGRAPHY

The number of papers on the fundamental theory of the many-body problem and its applications is vast. An attempt has been made to include references to most of the pioneering papers on the diagrammatic approach as well as a few others at appropriate places in these lecture notes. It might be useful here to list a few books and review papers covering other approaches and applications that go considerably beyond what we have presented here. The reader should find the present notes sufficient preparation for useful study in these further references.

Lecture Series, Symposia and Reprint Collections:

E. R. Caianello Editor, *Lectures on The Many Body Problem*, Academic Press, New York and London. 1962. (Lecture Notes from the Second International Spring School of the Istituto di Fisica Teorica of the Università di Napoli).

C. DeWitt, (Editor), *et al. The Many Body Problem*, J. Wiley & Sons, New York, 1958. (Les Houches lectures, 1958.)

C. Fronsdal, Editor, *The Many Body Problem*, W. A. Benjamin, New York, 1962. (Lecture Notes from the 1st Bergen International School of Physics—1961.)

D. Pines, *The Many Body Problem*, W. A. Benjamin, New York, 1961. (Reprint collection with extensive introduction and bibliography.)

D. Pines, *Elementary Excitations in Solids*, W. A. Benjamin, New York, 1963. (Lecture Series.)

H. L. Morrison, Editor, *The Quantum Theory of Many Particle Systems*, Gordon and Breach, New York and London, 1962. (Reprint collection.)

International Congress on Many-Body Problems, *Suppl. Physica* **26**, December, 1960 (contributed papers).

Review Articles and Books:

A. I. Alekseev, *Sov. Phys. Usp.* **4**, 23 (1961). "The Application of the Methods of Quantum Field Theory in Statistical Physics."

G. Baym and L. Kadanoff, *Quantum Statistical Mechanics*, W. A. Benjamin, New York, 1962.

V. L. Bonch-Bruevich and S. V. Tyablikov, *The Green Function Method in Statistical Mechanics*, North-Holland Publishing Company, Amsterdam, 1962.

G. V. Chester, *Advances in Phys.* **10,** 357 (1961). "The Theory of the Interaction of Electrons with Lattice Vibrations in Metals."

N. Fukuda and Y. Wada, *Prog. Theor. Phys. Suppl.* **15,** 61 (1960). "Some Aspects of Many Body Problems."

T. Kato, T. Kobayashi, and M. Namiki, *Prog. Theor. Phys. Suppl.* **15,** 3 (1960). "Formal Theory of Green Functions."

Yu. V. Novozhilov and A. V. Tulub, *The Method of Functionals in the Quantum Theory of Fields*, Gordon and Breach, New York, 1961.

D. J. Thouless, *Quantum Mechanics of Many-Body Systems*, Academic Press, New York and London, 1961.

D. N. Zubarev, *Sov. Phys. Usp.* **3**, 320 (1960). "Double-Time Green Functions in Statistical Physics."